The Origin of Higgs (“God”) Particles and the Higgs Mechanism:

Physics is Logic III, Beyond Higgs – A Revamped Theory With a Local Arrow of Time, The Theory of Everything Enhanced, Why Inertial Frames are Special, Universes of the Mind

STEPHEN BLAHA

BLAHA RESEARCH

ISBN: 978-0-9893826-9-4

Cover Credits

Rev. 00/00/01 November 3, 2015

To My Wife Margaret

Some Other Books by Stephen Blaha

All the Megaverse! Starships Exploring the Endless Universes of the Cosmos using the Baryonic Force (Blaha Research, Auburn, NH, 2014)

All the Universe! Faster Than Light Tachyon Quark Starships & Particle Accelerators with the LHC as a Prototype Starship Drive Scientific Edition (Pingree-Hill Publishing, Auburn, NH, 2011).

From Asynchronous Logic to The Standard Model to Superflight to the Stars (Blaha Research, Auburn, NH, 2011)

From Asynchronous Logic to The Standard Model to Superflight to the Stars; Volume 2: Superluminal CP and CPT, U(4) Complex General Relativity and The Standard Model, Complex Vierbein General Relativity, Kinetic Theory, Thermodynamics (Blaha Research, Auburn, NH, 2012)

The Algebra of Thought & Reality: The Mathematical Basis for Plato's Theory of Ideas, and Reality Extended to Include A Priori Observers and Space-Time; Second Edition (Pingree-Hill Publishing, Auburn, NH, 2009)

Quantum Big Bang Cosmology: Complex Space-time General Relativity, Quantum Coordinates, Dodecahedral Universe, Inflation, and New Spin 0, ½, 1 & 2 Tachyons & Imagyons™ (Pingree-Hill Publishing, Auburn, NH, 2004)

SuperCivilizations: Civilizations as Superorganisms (McMann-Fisher Publishing, Auburn, NH, 2010)

Standard Model Symmetries, And Four and Sixteen Dimension Complex Relativity; The Origin Of Higgs Mass Terms (Blaha Research, Auburn, NH, 2012)

The Bridge to Dark Matter; A New Sister Universe; Dark Energy; Inflatons; Quantum Big Bang; Superluminal Physics; An Extended Standard Model Based on Geometry (Blaha Research, Auburn, NH, 2013)

Universes and Megaverses: From a New Standard Model to a Physical Megaverse; The Big Bang; Our Sister Universe's Wormhole; Origin of the Cosmological Constant, Spatial Asymmetry of the Universe, and its Web of Galaxies; A Baryonic Field between Universes and Particles; Flatverse Extended Wheeler-DeWitt Equation (Blaha Research, Auburn, NH, 2014)

Available on bn.com, Amazon.com, Amazon.co.uk and other international web sites as well as at better bookstores (through Ingram Distributors).

Preface

This book points out shortcomings in the Higgs Mechanism and suggests a new mechanism. In doing so it shows that the origin of the Higgs particles is intimately associated with the gauge fields to which it gives masses.

In addition a new form of Higgs field is proposed that has retarded asymptotic propagators. We suggest these propagators are the *local* source of the Arrow of Time. All matter has mass due to this mechanism. All matter interacts with retarded Higgs fields. The cumulative result is a macroscopic Arrow of Time.

In addition, massive Higgs particles have a unique preferred rest frame. All inertial reference frames are obtained from it via Lorentz transformations. The result is that inertial reference frames have a preferred status in Special Relativity.

Using the mass generation mechanism presented in this book we therefore find 1) The origin of Higgs particles is the complex gauge fields made real-valued in The Extended Standard Model, except for the strong interactions whose gauge fields are necessarily complex in The Extended Standard Model; 2) Particle mass generation occurs through the choice of a specific vacuum state for Higgs particles; 3) A local Arrow of Time exists which becomes the global Arrow of Time due to the ubiquitous appearance of Higgs particle contributions in all matter; 4) Inertial reference frames are preferred due to the non-zero mass of Higgs particles – the Higgs particles rest frame is the frame from which all inertial reference frames obtain their special significance. These results strongly support the use of the new mass generation mechanism.

The new scalar particle theory shows that mass, the preferred inertial reference frames of Special Relativity, the Arrow of Time, and time itself are intimately related. It also shows a clear origin in complex gauge fields so that the "Higgs" sector is not a distinct addition to The Extended Standard Model but clearly intertwined with it.

CONTENTS

1. The Enigma of Higgs Particles and the Higgs Mechanism

In our previous work on the Standard Model, and its generalization to The Extended Standard Model described in a series of books culminating in two books, entitled *Physics is Logic ...*, we showed that the fermion spectrum results from Complex Special Relativity, the gauge interactions result from the Reality group, the fermion generations result from the Generation group, and the Theory of Everything results from the combination with Complex General Relativity. The Higgs particles and the Higgs Mechanism were inserted to generate particle masses and symmetry breaking effects.

Whence comes Higgs particles? There does not appear to be a more fundamental cause. And so the Higgs sector is an expedient mechanism to insert much needed symmetry breaking and masses into the theory.

There are a number of peculiarities in the implementation of the Higgs Mechanism:

1. First, it is selective in the sense that some gauge fields have associated Higgs particles and utilize the Higgs Mechanism, and some gauge fields do not have associated Higgs particles. In particular, the ElectroWeak gauge fields, the Generation group gauge fields, and the complex gravity field have associated Higgs particles. The strong interaction (gluon) gauge fields do not.

2. The Higgs potentials have a quadratic mass term of the "wrong" sign plus a quartic interaction term, which together, generate non-zero vacuum expectation values. They obviously accomplish their goal. But the source of these potentials, and why they have the same form, is unknown. One expects a fundamental principle should be operative here.

3. One can imagine creating a Higgs microscope at some super-accelerator. Using this microscope in the presence of a (classical) condensate could enable the Uncertainty Principle to be violated. This possibility, in the case of a microscope using electromagnetic fields, was the source of a heuristic argument for the need to quantize the electromagnetic field.[1]

4. The formulation of the Higgs Mechanism uses classical fields under the assumption that a path integral formulation justifies their use. While this may be true, the path integral formulation relies on implicit, unstated boundary conditions that obscure the physics of the quantum field theoretic nature of the mechanism. A direct

[1] Heitler (1954) p. 86 provides a good discussion of the need to quantize the electromagnetic field.

quantum field theoretic study of the Higgs Mechanism is needed and would further elucidate its character.

5. Scalar fields have a cloud hanging over them that spin ½ fields do not. A spin ½ particle cannot transition to negative energy because there is a filled sea of negative energy particles. No additional particles can fall into the sea due to the Pauli Exclusion Principle that forbids two fermions with the same 4-momentum and quantum numbers. In the case of scalar particles the Pauli Exclusion Principle does not apply and so a *filled* negative energy sea of scalar particles is not possible and positive energy scalar particles can transition to negative energy without hindrance. This problem has been "resolved" by an appropriate definition of the scalar particle vacuum to exclude transitions to negative energy. But the rationale for the definition is lacking. Dirac was asked about this issue many years ago. He said he had a solution to the problem. However he did not present it – in keeping with his well-known taciturn nature. So the issue remains an open question.

For the above reasons we will show that a more satisfactory method of achieving the goals of mass generation and symmetry breaking exists.[2] This method relies on a larger Fock space that enables the appearance of a vacuum expectation value for Higgs particles to be understood within a truly quantum framework. One major consequence of this approach is the appearance of a local Arrow of Time – a concept that has been a subject of interest for over one hundred years. Another consequence is a reason for ElectroWeak Higgs bosons and for their absence for the strong (gluon) interaction.

---

[2] In the Extended Standard Model we have shown that the basic iota particles have a mass, the Landauer mass, so that the theory is symmetry violating from the very start. We have also shown that our Two-Tier formalism for quantum field theories always yields finite results in perturbation theory calculations – making the renormalization approach of t'Hooft and others, which relied on an initially massless theory, unnecessary.

2. True Origin of an Acceptable Mass Creation Mechanism

The origin of the masses of Standard Model particles has been attributed to Higgs particles and the Higgs Mechanism. The apparent recent discovery of Higgs particles at CERN seems to solidify the Higgs sector of the Standard Model and of our Extended Standard Model as described in volumes I and II of *Physics is Logic*.[3]

However in view of the questions raised in the previous chapter we will now consider the possibility that the formulation of the Higgs sector should be reconsidered. In particular, point 1 of chapter 1 raises the question of why some gauge fields have associated Higgs particles and experience the Higgs Mechanism for mass generation and others do not. Turning the question around one might ask whether gauge fields are inherently associated with Higgs particles with some exceptions. This possibility has some support when one considers the new Higgs particles that we associate with the Generation group gauge fields and with Complex General Relativity's tensor gauge field in our Theory of Everything. (See Blaha (2015a) and (2015b).)

2.1 The Difference between the Strong Gauge Field and the Other Gauge Fields in the Extended Standard Model

In our Extended Standard Model the only gauge field without an associated Higgs particle is the strong interaction gluon gauge field. *We view this exception as a particularly important clue as to the nature of the relation between gauge fields and Higgs particles.*

How does the strong interaction gauge field differ from all other gauge fields in the Extended Standard Model and our Theory of Everything? An examination of the gauge fields dynamic equations (and lagrangian terms) of the Extended Standard Model reveals that all gauge field dynamic equation kinetic terms *except the strong interaction gauge field* have the form:

$$\partial/\partial x_\mu \, F^a{}_{\mu\nu} + gf^{abc} A^{b\mu} \, F^c{}_{\mu\nu} = j^a{}_\nu \tag{2.1}$$

where

$$F^a{}_{\mu\nu} = \partial/\partial x^\nu A^a{}_\mu - \partial/\partial x^\mu A^a{}_\nu + gf^{abc} A^b{}_\mu A^c{}_\nu \tag{2.2}$$

where a, b, c are structure constant indices, g is a coupling constant, and $j^a{}_\nu$ is the corresponding current. The gauge field $A^a{}_\mu$ is real for ElectroWeak gauge fields, and Generation group gauge fields. Thus eqns. 2.1 and 2.2 are real-valued.

---

[3] Blaha (2015a) and (2015b).

The strong interaction gauge field[4] differs from the other gauge fields by being *necessarily* complex due to the complex 3-space derivatives that appear in the corresponding equations:

$$D^{\mu} F_{C}{}^{a}{}_{\mu\nu} + gf^{abc} A_{C}{}^{b\mu} F_{C}{}^{c}{}_{\mu\nu} = j^{a}{}_{\nu} \tag{2.3a}$$

with

$$F_{C}{}^{a}{}_{\mu\nu} = D_{\nu} A_{C}{}^{a}{}_{\mu} - D_{\mu} A_{C}{}^{a}{}_{\nu} + gf^{abc} A_{C}{}^{b}{}_{\mu} A_{C}{}^{c}{}_{\nu} \tag{2.3b}$$

$$D_k = \partial/\partial x_r{}^k + i\, \partial/\partial x_i{}^k \qquad D_0 = \partial/\partial x^0 \tag{2.4}$$

for k = 1, 2, 3 where $A_{C}{}^{a}{}_{\mu}$ is a complexon gauge field. The complex spatial coordinate is $x_r{}^k + i\, x_i$. The time coordinate is real-valued. These equations are eq. 12.16 and 5.162 of Blaha (2015a) for complexon gauge fields,[5] which carry the strong interaction in the Extended Standard Model. Eq. 2.3a requires a complexon gauge field to be complex.

This difference enables us to differentiate the strong gauge field from the other gauge fields in The Extended Standard and thereby to develop a united formalism for the non-strong gauge fields and their corresponding Higgs particles.

2.2 The Genesis of Higgs Particle Fields from Complex Gauge Fields

In the prior section we considered the difference between the strong gauge field and the other gauge fields of The Extended Standard Model. Unlike strong gauge fields the other gauge fields (ElectroWeak and so on) could be real or complex. In a manner similar to what we did in the *Physics is Logic* books (and earlier books) we can assume the gauge fields are initially complex, and then transform them to real-valued fields using a phase transformation that introduces scalar fields that we will then take to be Higgs fields.

Since the gauge fields are transformable by Lorentz transformations we can assume each gauge field has a common phase for all its space-time components. Thus we define the phase transformation for a gauge field $A^{b\mu}$ by

$$A'^{a\mu}(x) = \Phi(x)^{a}{}_{b} A^{b\mu}(x) \tag{2.5}$$

where $\Phi(x) = \text{diag}(\exp[i\varphi_1(x)], \exp[i\varphi_2(x)], \ldots, \exp[i\varphi_n(x)])$, and n is the number of symmetry components of $A^{b\mu}$. Inserting $A'^{a\mu}(x)$ in eq. 2.1 we find that eq. 2.1 becomes:

$$\partial/\partial x_{\mu} F'^{a}{}_{\mu\nu} + gf^{abc} A'^{b\mu} F'^{c}{}_{\mu\nu} = j^{a}{}_{\nu} \tag{2.6}$$

where

---

[4] This field is called a complexon gauge field in Blaha (2015a) and earlier books.

[5] In The Extended Standard Model we also identify quark species particles as having complex 3-momentum. We call them complexon fermions.

$$F'^{a}_{\mu\nu} = \partial/\partial x^{\nu}\{\exp[i\varphi_a(x)]A^{a}_{\mu}\} - \partial/\partial x^{\mu}\{\exp[i\varphi_a(x)]A^{a}_{\nu}\} + gf^{abc}\exp[i\varphi_b(x)]\exp[i\varphi_c(x)]A^{b}_{\mu}A^{c}_{\nu} \quad (2.7)$$

If we now assume that $\varphi_a(x)$ is small for all a then

$$\exp[i\varphi_a(x)] \simeq 1 + i\varphi_a(x) \quad (2.8)$$

to first order. Substituting in eqs. 2.6 and 2.7 and keeping terms to leading order yields the real part:

$$\partial/\partial x_{\mu}\, F^{a}_{\mu\nu} + gf^{abc}A^{b\mu}\, F^{c}_{\mu\nu} = j^{a}_{\nu} \quad (2.9)$$

where $F^{a}_{\mu\nu}$ is given by eq. 2.2, and the imaginary part:

$$\partial/\partial x_{\mu}\, F_{i\,\mu\nu}^{a} + gf^{abc}A^{b\mu}\, F_{i\,\mu\nu}^{c} = 0 \quad (2.10)$$

to leading order where

$$F_{i\,\mu\nu}^{a} = \partial/\partial x^{\nu}\, \varphi_a(x)A^{a}_{\mu} - \partial/\partial x^{\mu}\, \varphi_a(x)A^{a}_{\nu} \quad (2.11)$$

Substituting eq. 2.11 in eq. 2.10 we find

$$A^{a}_{\nu}\Box\varphi_a(x) - A^{a}_{\mu}\, \partial/\partial x_{\mu}\partial/\partial x^{\nu}\, \varphi_a(x) - gf^{abc}A^{b\mu}\, [A^{c}_{\mu}\, \partial/\partial x^{\nu}\, \varphi_a(x) - A^{c}_{\nu}\, \partial/\partial x^{\mu}\, \varphi_a(x)] = 0 \quad (2.12)$$

in the Landau gauge, with no sum over a. Eq. 2.12 is a form of Klein-Gordon equation having interaction terms with the gauge field. If the gauge field is weak then only the first two terms are important.

Note that only derivatives of $\varphi_a(x)$ appear in eq. 2.12. Consequently shifts of the $\varphi_a(x)$ field by a constant yield solutions of eq. 2.12. This feature makes $\varphi_a(x)$ a candidate to be a Higgs particle.

Note also that complexon fields cannot have a phase change due to the complexity of the spatial coordinates. This difference appears to be the reason why the strong interaction gauge field does not have an associated Higgs particle.

The $\varphi_a(x)$ particles can be made into Higgs particles by adding an appropriate potential:

$$V= A\, \varphi_a^{2}(x) + B\, \varphi_a^{4}(x) \quad (2.13)$$

where A and B are constants. Approximating eq. 2.12 with its first two terms and inserting the potential term we find the Higgs-like equation:

$$A^{a}_{\nu}\Box\varphi_a(x) - A^{a}_{\mu}\, \partial/\partial x_{\mu}\partial/\partial x^{\nu}\, \varphi_a(x) + \partial V/\partial\varphi_a = 0 \quad (2.14)$$

$\varphi_a(x)$has a minimum at the minimum of the potential in the coresponding lagrangian.

The second and third terms in eq. 2.14 constitute the interaction. Neglecting these terms we see that eq. 2.14 becomes the free, massless, field Klein-Gordon equation

$$\Box\varphi_a(x) = 0 \tag{2.15}$$

We will not pursue the Higgs Mechanism approach here but will instead develop an alternate approach to generate particle masses in the following chapters.

The pairing of Higgs particles with real-valued gauge fields is thus established. The non-existence of a matching Higgs field for the strong interaction is due to the inherently complex nature of the strong interaction (complexon) gauge field in the Extended Standard Model also follows.

The derivation presented here is analogous to the derivation of Higgs fields in Complex General Relativity – also a gauge theory – in *Physics is Logic Part II*.

One of the remarkable aspects of The Extended Standard Model is its ability to directly determine qualitative properties of elementary particles: four fermion species, Parity violation, the distinction between leptons and quarks, the match of the Standard Models (broken) symmetries with the Reality group consisting of subgroups of U(4), and now the existence of Higgs gauge fields in the ElectroWeak sector but not for the strong interactions. We take these successes to be indicators of the correctness of The Extended Standard Model.

3. A New Mass Creation Mechanism

The Higgs Mechanism, which is based on the introduction of c-number vacuum values of Higgs fields directly in the lagrangian, may be technically correct within a path integral formulation. But, in the author's opinion, the path integral formulation, which is based on implicit boundary conditions, masks some of the essential features of spontaneous breakdown using Higgs particles.

In this chapter we will describe a new mechanism that utilizes an extension of quantum field theory to include classical fields that we have called *pseudoquantization.*[6] It combines both quantum and classical fields within the same framework. In this extended theory vacuum expectation values appear as coherent ground states that are strictly classical in nature.

This chapter will be based on our 1978 paper that appeared in the peer-reviewed journal *Physical Review D.* The paper is reproduced in the Appendix for the reader's convenience with the kind permission of The American Physical Society.

We suggest the reader skim or read the paper before proceeding, or refer to the paper for details as necessary while reading the chapter. The paper also does present a new formulation of Quantum Mechanics that incorporates both quantum and classical mechanics within one framework. Recently, experimenters have been investigating the possibility of macroscopic quantum phenomena. The new formulation is ideally suited for tracing the change from a quantum to a classical regime. It also is applicable to "large n atoms" where the outermost electrons approach classical behavior with an almost continuous energy spectrum.

3.1 Pseudoquantization of Higgs Particles

We will now consider the pseudoquantization of a scalar particle field that will become a Higgs particle with a non-zero vacuum expectation value. This process was presented in section III of our paper in the Appendix. We begin by defining two fields that correspond to the scalar particle: $\varphi_1(x)$ and $\varphi_2(x)$.[7] These fields will be assumed to have the equal time commutators

$$
\begin{aligned}
[\varphi_i(x), \pi_j(y)] &= i(1 - \delta_{ij})\delta^3(\mathbf{x} - \mathbf{y}) \qquad (3.1)\\
[\varphi_i(x), \varphi_j(y)] &= 0\\
[\pi_i(x), \pi_j(y)] &= 0
\end{aligned}
$$

where δ_{ij} is the Kronecker δ and where $\pi_i(x)$ is the canonically conjugate momentum to $\varphi_i(x)$. The fields $\varphi_1(x)$ and $\pi_1(y)$ will be observable classical fields as shown by eqns. 69 and 70 in the

---

[6] This new formalism was first described in S. Blaha, Phys. Rev. D**17**, 994 (1978).

[7] The subscripts on the fields are not gauge symmetry indices but simply identifiers distinguishing the fields from each other.

Appendix. The fields $\varphi_2(x)$ and $\pi_2(y)$ will not be observables so that $\varphi_1(x)$ and $\pi_1(y)$ can both be sharp on the set of physical states.

We now specify the lagrangian density for a scalar Klein-Gordon particle:

$$\mathcal{L} = \partial\varphi_1/\partial x_\mu \partial\varphi_2/\partial x^\mu \quad (3.2a)$$

with hamiltonian density

$$\mathcal{H} = \pi_1\ \pi_2\ + \partial\varphi_1/\partial x_i \partial\varphi_2/\partial x^i \quad (3.2b)$$

where i labels spatial coordinates, and $\pi_1 = \partial\varphi_2/\partial t$ and $\pi_2 = \partial\varphi_1/\partial t$. Eqs. 3.2 are without a potential or mass term. They correspond to eq. 2.15.

The lagrangian and hamiltonian for a massive particle are

$$\mathcal{L} = \partial\varphi_1/\partial x_\mu \partial\varphi_2/\partial x^\mu - m^2\ \varphi_1\varphi_2 \quad (3.2c)$$

with hamiltonian density

$$\mathcal{H} = \pi_1\ \pi_2\ + \partial\varphi_1/\partial x_i \partial\varphi_2/\partial x^i + m^2\ \varphi_1\varphi_2 \quad (3.2d)$$

The fields can be fourier expanded in terms of creation and annihilation operators:

$$\varphi_i(\mathbf{x}, t) = \int d^3k\ [a_i(k)f_k(x) + \ a_i^\dagger(k)f_k{}^*(x)] \quad (3.3)$$

for i = 1, 2 where

$$f_k(x) = e^{-ik\bullet x}\ /(2\omega_k(2\pi)^3)^{1/2}$$

with $\omega_k = |\mathbf{k}|$.

The creation and annihilation operators satisfy the commutation relations:

$$[a_i(k),\ a_j^\dagger(k')] = (1 - \delta_{ij})\delta^3(\mathbf{k} - \mathbf{k'}) \quad (3.4)$$
$$[a_i(k),\ a_j(k')] = 0$$
$$[a_i^\dagger(k),\ a_j^\dagger(k')] = 0$$

for i, j = 1, 2.

In this formulation the defining properties of a physical state are:

$$\varphi_1(x)|\Phi, \Pi> = \Phi(x)|\Phi, \Pi> \quad (3.5)$$
$$\pi_1(x)|\Phi, \Pi> = \Pi(x)|\Phi, \Pi>$$

where $\Phi(x)$ and $\Pi(x)$ are sharp on the states and thus classical fields with

$$\Phi(\mathbf{x}, t) = \int d^3k\ [\alpha(k)f_k(x) + \ \alpha^*(k)f_k{}^*(x)] \quad (3.6)$$

and correspondingly for $\Pi(x)$.

3.2 Vacuum States for Scalar Particles with Non-Zero Vacuum Expectation Values

When we implement the mass mechanism Φ becomes constant. We can define a set of states

$$a_1(k)|\alpha> = \alpha(k)|\alpha>$$
$$a_1^\dagger(k)|\alpha> = \alpha^*(k)|\alpha>$$

and correspondingly a set of coherent states

$$|\alpha> = C\exp\{\int d^3k\,[\alpha(k)a_2^\dagger(k) + \alpha^*(k)a_2(k)]\}|0> \qquad (3.7)$$

where C is a normalization constant and where the vacuum state |0> satisfies

$$a_1(k)|0> = a_1^\dagger(k)|0> = 0 \qquad (3.8a)$$
$$a_2(k)|0> \neq 0 \qquad\qquad a_2^\dagger(k)|0> \neq 0 \qquad (3.8b)$$

The dual vacuum state satisfies

$$<0|a_2(k) = <0|a_2^\dagger(k) = 0 \qquad (3.9a)$$
$$<0|a_1(k) \neq 0 \qquad\qquad <0|a_1^\dagger(k) \neq 0 \qquad (3.9b)$$

Additional details on these coherent states, which differ from conventional coherent states such as those of Kibble and others can be found in the Appendix.

With this coherent state formalism, which gives purely classical fields and yet also has quantum fields through the use of φ_2 and its creation and annihilation operators, we now have the machinery to define a mass mechanism without the introduction of a potential whose origin can only be described as dubious.

For we can define a coherent state:

$$|\Phi, \Pi> = C\exp\{[(2\pi)^3\omega_k/2]^{1/2}\Phi[a_2^\dagger + a_2]\}|0> \qquad (3.10)$$

where C is a normalization constant, that yields a non-zero vacuum expectation value:

$$\varphi_1(x)|\Phi, \Pi> = \Phi|\,\Phi, \Pi> \qquad (3.11)$$

where Φ is a constant. Evaluating a fermion interaction term we find a mass term emerges[8]

$$\bar{\psi}(\varphi_1 + \varphi_2)\psi \rightarrow \bar{\psi}(\Phi + \varphi_2)\psi \qquad (3.12)$$

and it generates a mass for an interaction with a gauge field of the form

---

[8] When matrix elements with a "vacuum state" such as eq. 3.10 are taken.

$$A^{\mu}(\varphi_1 + \varphi_2)^2 A_{\mu} \rightarrow A^{\mu}(\Phi + \varphi_2)^2 A_{\mu} \quad (3.13)$$

It also yields a quantum field theoretic interaction that would result in the production of ElectroWeak particles from these scalar fields. The production of Higgs particles that decay into ElectroWeak gauge particles has recently been found at CERN.

The present formalism provides a clean way to separate the vacuum expectation value of a scalar particle from its quantum field part in contrast to the Higgs Mechanism where one has to separate a Higgs field into parts manually.

3.3 Interpretation of Negative Energy Scalar Particle States

As we noted in chapter 1 scalar particle physics has the problem of no barrier to the decay of positive energy states to negative energy states due to the absence of a Pauli Exclusion Principle for bosons. The pseudoquantization procedure that we developed in 1978 and describe here allows negative energy states as one would physically expect and raises the possibility of disastrous particle decays to negative energy. Eqs. 3.7 and 3.8 show that negative energy states are possible in this theory.

However eq. 3.7 also shows that combined positive and negative energy boson states can be interpreted as classical field states. In addition the ability of any number of boson particles to have the same 4-momentum and quantum numbers shows that a *macroscopic* classical scalar field state can be constructed.

Thus we can view states containing negative energy particles as classical field states and thus solve the issue of interpreting negative energy particle states – a more satisfactory approach than the standard quantization procedure does – with due respect to Professor Dirac.

We note that macroscopic many particle fermion states can only have one particle in any mode unlike bosons. Therefore we cannot use this formalism to create macroscopic classical fermion field states.

3.4 Contrast with Conventional Second Quantization of Scalar Particles

The pseudoquantization procedure followed in this chapter uses different boundary conditions than the usual scalar particle quantization procedure. The essence of the difference is embodied in a comparison of the definition of the vacuum in eqs. 3.8 and 3.9 and the definition of the conventional second quantized field vacuum:

$$a|0\rangle = 0 \qquad \text{Conventional Approach} \qquad (3.14)$$
$$a^{\dagger}|0\rangle \neq 0$$

In the conventional approach the creation of negative energy boson states is eliminated *ab initio* whereas in our approach it is allowed in order to support classical field states with non-zero vacuum expectation values that are a form of classical field. While one cannot discredit the conventional choice for conventional scalar fields, one can see that our approach yields a physically more important result – particularly for Higgs fields – because it leads to the arrow of time *locally* – an important feature of physical phenomena that has been a subject of much discussion and dispute. One can say that the conventional approach sweeps the issue "under the rug" rather than seeking a deeper justification – differing from Dirac's implied notion that the issue merited attention. We will discuss the "arrow of time" within the framework of our pseudoquantization approach later.

3.5 Why Inertial Reference Frames are Special

The great physicists of the early 20th century raised numerous questions about Special Relativity after Einstein and Poincarè's discovery. Prominent among them was the question of why inertial reference frames are of especial importance in Special Relativity, and afterwards in General Relativity.

It appears that our formulation of the mass generation mechanism sheds significant light on the reason for the special prominence of inertial frames. Earlier we considered the case of a massless pseudoquantized scalar. We now consider massive scalars since experiments at CERN have apparently discovered a Higgs particle with a 125 GeV/c mass. Eqs. 3.2c and 3.2d describe a massive scalar particle. If the scalar is massive, then the "vacuum" state, eq. 3.10, that yields a non-zero expectation value must change to

$$|\Phi, \Pi> = C\exp\{(2\pi)^3 m/2]^{\frac{1}{2}}[a_2^\dagger(\mathbf{0},m) + a_2(\mathbf{0},m)]/2\}|0> \qquad (3.10')$$

to have operators for a particle of mass m in its rest frame. Then, having established this preferred frame for a Higgs particle, in The Extended Standard Model, and requiring that invariant intervals

$$ds^2 = dt^2 - d\mathbf{x}^2 \quad \text{(in rectangular coordinates)} \qquad (3.15)$$

are unchanged by a (complex or real) Lorentz transformation, we find that inertial reference frames are singled out as "special" in the sense that they are the only accessible reference frames that can be generated by a Lorentz boost/transformation from the Higgs particle rest frame. *The Higgs particle vacuum state singles out the class of inertial reference frames.*

Thus Higgs particles play a central role in establishing the basis of physical reality.

3.6 Pseudoquantization Reveals More Physical Consequences of Higgs Particles

Earlier we pointed out that our pseudoquantization theory of Higgs particles reveals more physical consequences than the conventional approach, which implements the Higgs Mechanism by simply using a potential term that has a minimum at a non-zero vacuum expectation value. This chapter and the following chapters show the major results of a properly implemented mechanism. We find a better explanation of the negative energy state problem of boson field theories. We find a local arrow of time that explains the direction of time that we, and all of nature, experiences. We find the reason why inertial reference frames have a special physical significance – a result long sought by physicists.

In addition we saw that real gauge fields should have an associated Higgs particle, while necessarily complex gauge fields (the strong interaction gauge field in The Extended Standard Model) do not have an associated gauge field. These results correspond to experimental reality.

4. T Invariance of Our Quantized Scalar Particle Theory

The pseudoquantized scalar particle hamiltonian equations are invariant under time reversal t → t' = –t. The vacuum states defined by eqs. Eqs. 3.8 and 3.9 break the time reversal invariance of the theory resulting in retarded particle propagators.

The hamiltonian equations

$$[H, \varphi_1(\mathbf{x}, t)] = -i\partial\varphi_1/\partial t \qquad (4.1)$$
$$[H, \varphi_2(\mathbf{x}, t)] = -i\partial\varphi_2/\partial t$$

are invariant under time reversal. If we define a time reversal operator transformation U then the time reversed equations are

$$[UHU^{-1}, \varphi_1(\mathbf{x}, -t)] = +i\partial\varphi_1(\mathbf{x}, -t)/\partial(-t) \qquad (4.2)$$
$$[UHU^{-1}, \varphi_2(\mathbf{x}, -t)] = +i\partial\varphi_2(\mathbf{x}, -t)/\partial(-t)$$

The operator U, which is unitary, transforms H into –H. This operation is legal because the hamiltonian – in this case – is not positive definite and admits negative energy states.[9] Thus

$$[H, \varphi_1(\mathbf{x}, -t)] = -i\partial\varphi_1(\mathbf{x}, -t)/\partial(-t) \qquad (4.3)$$
$$[H, \varphi_2(\mathbf{x}, -t)] = -i\partial\varphi_2(\mathbf{x}, -t)/\partial(-t)$$

and the time reversal invariance of the equations of motion is established for this case.

Time reversal invariance is broken by our choice of vacuum states. This choice is necessary to obtain classical field states as we showed in the preceding chapter and in the Appendix. A demonstration of the time reversal symmetry breaking is presented in the following chapter where we show theory has retarded propagators for particle propagation to and from asymptotic states.

Within the interaction region the particle propagators are the sum of retarded and advanced parts that combine to yield principle value propagators – not Feynman propagators. Many years ago Feynman and Wheeler championed principle value propagators for electrodynamics to obtain an action-at-a distance theory of Quantum Electrodynamics. While their theory, and ours, differ from the standard quantum field theory approach there is no reason to view them as faulty, or having serious physical defects. The only question is whether nature chooses conventional quantum field theory or pseudoquantized quantum field theory. In our case the need for a classical scalar particle non-zero vacuum expectation strongly motivates our choice of psedoquantized Higgs particles.

---

[9] Unlike the usual case of second quantized Klein-Gordon quantum field theory.

5. Retarded Propagators for Our Quantized Higgs Particles

In the previous chapter we pointed out that our pseudoquantization Higgs theory has an arrow of time due to is boundary conditions as expressed by its definition of the vacuum state and its dual. In this chapter we will show that the theory uses retarded propagators for propagation to and from the interaction region to asymptotic in-states and out-states. Within an interaction region the theory uses half-retarded – half-advanced propagators. The Appendix paper has a detailed discussion of propagators between eqs. 145 and 149, and also in section V. We will discuss aspects of the perturbation theory and propagators of our scalar particles in this chapter.

First we note that in-states at $t = -\infty$ are composed of superpositions of $a_2(k)$ and $a_2^\dagger(k)$ creation and annihilation operators by eq. 3.8b:

$$a_2(k)|0\rangle \neq 0 \qquad\qquad a_2^\dagger(k)|0\rangle \neq 0 \tag{3.8b}$$

while the out-states composed of superpositions of $a_1(k)$ and $a_1^\dagger(k)$ creation and annihilation operators by eq. 3.9b:

$$\langle 0|a_1(k) \neq 0 \qquad\qquad \langle 0|a_1^\dagger(k) \neq 0 \tag{3.9b}$$

Consequently when in-state particles (x_1) propagate into the interaction region (x_2) the relevant propagators are retarded propagators with the form

$$\begin{aligned} G_{in}(x_2, x_1) &= \langle 0|T(\varphi_{1\,in}(x_2), \varphi_{2\,in}(x_1))|0\rangle \\ &= \theta(x_{20} - x_{10})\langle 0|[\varphi_{1\,in}(x_2), \varphi_{2\,in}(x_1)]\,|0\rangle \end{aligned} \tag{5.1}$$

by eq. 148 of the Appendix. Eq. 5.1 is a manifestly retarded propagator. The choice of vacuums clearly results in a time asymmetry giving a retarded propagation reflecting the familiar Arrow of Time.

A similar situation prevails for propagation to out-states (x_3) from the interaction (x_2) region:

$$\begin{aligned} G_{out}(x_3, x_2) &= \langle 0|T(\varphi_{1\,out}(x_3), \varphi_{2\,out}(x_2))|0\rangle \\ &= \theta(x_{30} - x_{20})\langle 0|[\varphi_{1\,out}(x_3), \varphi_{2\,out}(x_2)]\,|0\rangle \end{aligned} \tag{5.2}$$

Within the interaction region the Higgs particles have principle value propagators as shown in section V of the Appendix,

Thus we find pseudoquantized Higgs particles embody a local Arrow of Time. The locality of the Arrow of Time is embodied in all the particles that interact with the Higgs particle. Since the mass of *every* particle – bosons and fermions – has a Higgs contribution, and thus *every* particle interacts with the Higgs particles, the Arrow of Time permeates The Extended Standard Model as well as the more familiar Standard Model known from experiment.

6. The *Local* Arrow of Time

In the *Physics is Logic* monographs we saw that complex coordinates led to the form of the fermion spectrum, that the mapping of complex coordinates to real-valued coordinates yielded the Reality group and The Extended Standard Model gauge interaction, that Complex General Relativity led to Higgs particles that were directly united with elementary article masses and gave us the equality of inertial mass and gravitational mass, and, in the present volume, that the reduction of complex gauge fields to real gauge fields explained the appearance of Higgs fields in The Standard Model and The Extended Standard Model.

Now we see that the pseudoquantization procedure leads to retarded Higgs field propagators and thence to a *local* arrow of time. Many arguments have been put forward over the past hundred plus years for the Arrow of Time. Many arguments based on Statistical Mechanics, Entropy, and Boltzmann's statistical atomic theory have suggested the Arrow of Time is a global statistical consequence. This view seems to contradict the results of elementary particle experiments where a *local* Arrow of Time is evident.

Our rationale for the Arrow of Time begins with retarded Higgs fields. Then we note that Higgs field quantum interactions appear for all fermions and gauge particles. Thus all particle interactions are imbued with an Arrow of Time. Particles united to form macroscopic matter inherit their combined Arrows of Time producing the global Arrow of Time we experience.

Thus our pseudoquantiztion approach offers a more satisfactory solution of the origin of the Arrow of Time.

It is remarkable that complex quantities – coordinates and fields – through the Higgs phenomena we have considered lead to the equality of inertial mass and gravitational mass, and an Arrow of Time. This unity of mass and time phenomena may reflect the deeper fact that we can have no practical Arrow of Time if all particles were massless, for particle dynamics at light speed would then be pointless. This view has been expressed by DeWitt, Unruh, and others who have pointed out that, physically, time is meaningful and measurable only if masses exist; the larger the mass, the more accurate the time measurement in principle.[10]

[10] No mass, no clock; no clock, no physical time. See Blaha (2015a) pp. 368-371 for a discussion including comments by DeWitt and Unruh.

7. Theory of Everything Revisited

7.1 Previously Obtained Results

Our recent books, Blaha (2015a) and (2015b), have developed a Theory of Everything that contains:

1. The Extended Standard Model, which contains the experimentally known Standard Model as a subset.
2. Complex General Relativity, which contains Einstein's General Relativity as its real-valued limit.
3. A unification of these theories in a Theory of Everything based on the Reality Group and the use of the Higgs Mechanism.

We conclude

1. The core of every fermion is an *iota* that consists of four spin states (or logical values) and an ultra-small mass that we call the Landauer mass.
2. The coordinates of our universe are complex-valued.
3. The complexity of coordinate values is hidden because rulers and clocks only yield real values.
4. Theoretically complex-valued coordinates are mapped to real values by the Reality group in recognition of item 2.
5. The map leads to the gauge interactions of The Extended Standard Model.
6. Due to the existence of Higgs particles in both parts of the Theory of Everything, that contributes to all fermion masses we find inertial mass equals gravitational mass – an experimental fact whose origin has been a subject of speculation for hundreds of years.
7. Parity violation is a necessary part of The Extended Standard Model due to its basis in complex-valued space-time.

7.2 New Results in The Theory of Everything

In this work we show that it is no accident that most gauge fields are associated with Higgs bosons. If we assume gauge fields are initially complex, then Higgs bosons naturally arise from the use of a Higgs field to map the gauge boson fields to real values.[11] This approach can also be followed in Complex General Relativity. It yields a real-valued Einstein General Relativity theory and Higgs bosons that contribute to elementary particle masses and establishes the equality of inertial mass and gravitational mass.

We have thus shown that the origin of the Higgs sector is intimately associated with the gauge fields to which it gives masses. Noting the shortcomings of a Higgs Mechanism based on the use of a Higgs potential, we use pseudoquantization to produce a larger quantum field theory that incorporates both classical fields and quantum fields. The non-zero vacuum expectation of Higgs bosons then is a constant, classical field, as it should be, due to the choice of a vacuum state.

Pseudoquantized Higgs fields have retarded asymptotic propagators that we suggest are the *local* source of the Arrow of Time. All matter has mass due to this mechanism. All matter interacts with retarded Higgs fields. The cumulative result is a macroscopic Arrow of Time. We believe that the Arrow of Time must be local, or differing regions of the universe would have differing Arrows of Time, which would have been found in astronomical studies.

Lastly, in section 3.5 we pointed out that massive Higgs particles (CERN studies found a 125 GeV/c Higgs particle) have a unique rest frame that establishes a preferred reference frame. Physical reference frames must be inertial reference frames generated by Lorentz boosts and rotations from the Higgs rest frame.

All these results emanate from our derivation of Higgs particles from complex gauge fields made real, and the use of pseudoquantization for Higgs fields due to the classical nature of Higgs vacuum expectation values.

In summary we have found the following additional results IF the Higgs Mechanism sector of our Theory of Everything is changed to the mass generation mechanism presented in this book:

1. The origin of Higgs particles from all complex gauge fields except the strong interactions whose gauge fields are necessarily complex in The Extended Standard Model.
2. Particle mass generation by the choice of a specific vacuum state for Higgs particles.
3. The existence of a local Arrow of Time which becomes the global Arrow of Time due to the ubiquitous appearance of Higgs particle contributions in all matter.
4. The existence of a preferred reference frame due to the non-zero mass of Higgs particles – the Higgs particles vacuum expectation values common rest state.

We conclude by noting the very important role of complex-valued coordinates and complex-valued gauge fields in The Theory of Everything. Their transformation to real values lead

[11] The only exception in reality, and in our Extended Standard Model, are the strong interaction gauge fields, which are necessarily complex because their dynamic equations use complex 3-coordinates (complexon coordinates). Strong interaction gauge fields have no associated Higgs bosons.

directly to the detailed form of The Theory of Everything. In particular, they lead to the following possible symmetries:

1. Extended Standard Model: R⊗U(4) = SU(3)⊗SU(2)⊗U(1)⊗SU(2)⊗U(1)⊗U(4)
2. High energy GUT limit of The Extended Standard Model (if it exists) U(4)⊗U(4)
3. Theory of Everything: SU(3)⊗SU(2)⊗U(1)⊗SU(2)⊗U(1)⊗U(4)⊗U(4)
4. GUT limit of The Theory of Everything: U(4)⊗U(4)⊗U(4)
5. Possible Super-GUT limit of the Theory of Everything: SU(7)

We have suggested SU(7) as the possible Super-GUT limit of The Theory of Everything in anticipation of an extraordinary high energy limit of The Theory of Everything, in which The Extended Standard Model unites with General Relativity to form a regime where there is only one coupling constant. We note that the 48 generators of U(4)⊗U(4)⊗U(4) equals the 48 generators of SU(7).[12]

[12] Although U(4)⊗U(4)⊗U(4) is not a subgroup of SU(7).

8. Universes of the Mind

Having reached the point of a Theory of Everything, and being confident of its success because of its strong base in geometry and logic, one may ask is everything done and what remains to be done?

The glaring mystery in our formulation is the absence of a means to determine the vast number of "fundamental" constants that appear in the theory. The form follows fairly directly from geometry and logic. But the origin of the coupling constants and masses that appear in the theory is unknown. One could posit symmetry conditions that may determine some or, even, perhaps all of them. But we are unable to determine them from fundamental conditions.

The only recent quantum field theoretic attempt to determine a fundamental constant was the attempt of Johnson, Baker, and Willey to determine the fine structure constant α (approximately 1/137) in a model of massless QED.[13] Their effort seems to have failed.[14] And no further attempts have been made to calculate fundamental constants in 4-dimensional quantum field theories to the best of the author's knowledge.

Other than consistency conditions there does not appear to be a viable method to determine fundamental constants. So one can either assume we must work harder to find a yet deeper reason for the values of constants or we can assume that their values result from as yet unknown initial value conditions operative at the beginning of the universe – the Big Bang point.

In the absence of a determination of constants from fundamental considerations we can take these constants to be parameters and proceed, in future, to study universes with other values for the constants. Our proposal in Blaha (2015a), that our universe is but one of many universes existing in a 16-dimensional Megaverse, lends weight to the study of ""universe species." We can take a Darwinian viewpoint and ask of the evolution of other universes with different sets of fundamental constants. Some efforts in this direction have been made in considerations of the anthropomorphic conditions needed for life, as we know it, to evolve in a universe.

A more general study of universes with all sorts of fundamental constants under all sorts of conditions might shed light on the reason our universe has the constants it has. Often the differences between things provide clues as to why things are as they are. One might call this study, "Comparative Universe Studies."

---

[13] M. Baker and K. Johnson, Phys. Rev. **D8**, 1110 (1973) and references therein.

[14] S. Blaha, Phys. Rev. **D9**, 2246 (1974). This calculation tried to determine the fine structure constant by summing an infinite number of diagrams for the α eigenvalue function – perhaps the largest summation in 4-dimensional perturbation theory ever performed. The calculation agreed with known exact results to 4th order in α in perturbation theory. It found $\alpha = 1$ and it did not find an essential singularity which S. Adler had suggested would appear at the α eigenvalue point if there was a true solution. In comparison, many years earlier Heisenberg attempted to calculate α and found $\alpha = \pi$.

Perhaps a more interesting study of universes could simply be at the level of entities in the universe: the galaxies, the stars, the planets, and the life. This study could be scientific and probe issues such as the nature and possibilities for life forms. Or it could be "humane" and study forms of intelligence. This type of study, of course, runs the risk of being Science Fiction.

Lastly, one could imagine that a sufficiently intelligent, and mentally capacious, individual might be able to create a universe – perhaps a not very physical universe – in its mind and have the creatures in this universe undergo evolution and "live their lives" in a "historical" sequence within a civilization. Carried to an extreme this process could be viewed as the manner in which God works.

If the origin of the set of constants governing our universe is unknown, it behooves us to seek understanding by considering the space of all sets of constants with a view towards understanding the constants of our own universe.

Appendix. Pseudoquantization Theory Paper

This appendix has a reproduction of a paper by the author in 1978 that establishes an extended quantum field theory formalism for lagrangian theories that contain both classical and quantum fields. This paper (S. Blaha, Phys. Rev. D**17**, 994 (1978)) provides a basis for theories in which quantum fields have a non-zero vacuum expectation value that introduces a classical value into an otherwise quantum theory.

Embedding classical fields in quantum field theories

Stephen Blaha*
Physics Department, Syracuse University, Syracuse, New York 13210
(Received 2 August 1976; revised manuscript received 7 November 1977)

We describe a procedure for quantizing a classical field theory which is the field-theoretic analog of Sudarshan's method for embedding a classical-mechanical system in a quantum-mechanical system. The essence of the difference between our quantization procedure and Fock-space quantization lies in the choice of vacuum states. The key to our choice of vacuum is the procedure we outline for constructing Lagrangians which have gradient terms linear in the field variables from classical Lagrangians which have gradient terms which are quadratic in field variables. We apply this procedure to model electrodynamic field theories, Yang-Mills theories, and a vierbein model of gravity. In the case of electrodynamics models we find a formalism with a close similarity to the coherent-soft-photon-state formalism of QED. In addition, photons propagate to $t=+\infty$ via retarded propagators. We also show how to construct a quantum field for action-at-a-distance electrodynamics. In the Yang-Mills case we show that a previously suggested model for quark confinement necessarily has gluons with principal-value propagation which allows the model to be unitary despite the presence of higher-order-derivative field equations. In the vierbein-gravity model we show that our quantization procedure allows us to treat the classical and quantum parts of the metric field in a unified manner. We find a new perturbation scheme for quantum gravity as a result.

I. INTRODUCTION

The relation between classical and quantum systems has been a subject of continuing interest over the years: First, in the original development of quantum mechanics, second, in the study of the classical limit and infrared divergences of quantum-electrodynamic processes,[1,2] and third, in recent attempts to construct strong-interaction models of quark confinement which are for the most part either classical field theory models in search of quantization[3] or quantized gluon models wherein quark confinement is a consequence of infrared behavior.[4,5]

We will describe a new quantization procedure (called pseudoquantization) for field theory which is the analog of Sudarshan's method for embedding a classical-mechanical system in a quantum-mechanical system. It can be used with advantage to either embed a classical field theory in a quantum field theory in such a way as to maintain the classical character of the embedded fields (while studying the interaction between the classical and quantum sectors on essentially the same footing), or to quantize a class of field theories, members of which have been used as models for gravity and as models for the strong interaction with quark confinement.[7-9]

We shall begin (Sec. II) by pseudoquantizing a classical simple harmonic oscillator. This case is of particular importance because of the analogy between the mode amplitudes of a quantum field and the coordinates of a set of simple harmonic oscillators which we will take advantage of in later sections.

In Sec. III we describe the pseudoquantization procedure for field theory. We apply it to electrodynamic models and show that the propagation of photons to $t=+\infty$ is necessarily retarded in this formalism. Further, we display a close analogy between the present formalism and the coherent-soft-photon-state formalism[10] of QED.

In Sec. IV we apply the pseudoquantization procedure to a classical Yang-Mills field. The resulting field theory (with a slight but important modification) has been used as a model for the strong interactions with quark confinement.[7-9] We also apply the pseudoquantization procedure to a vierbein model of gravity and obtain a new perturbation theory for quantum gravity.

In Sec. V we show that principal-value propagators naturally arise in certains sectors of pseudoquantized theories thus verifying an *ad hoc* procedure devised to unitarize a model of quark confinement.[7-9] We also show how to construct a quantum version of action-at-a-distance electrodynamics.

We shall now briefly outline the procedure for embedding a classical-mechanical system in a quantum system.[6] Consider a classical Hamiltonian system with one degree of freedom, and commuting canonical variables, x_1 and p_1, which have the equations of motion

$$\dot{x}_1 = -i[x_1, \hat{H}], \tag{1}$$

$$\dot{p}_1 = -i[p_1, \hat{H}], \tag{2}$$

where defining

$$\hat{H} = -i\left(\frac{\partial H(x_1,p_1)}{\partial p_1}\frac{\partial}{\partial x_1} - \frac{\partial H(x_1,p_1)}{\partial x_1}\frac{\partial}{\partial p_1}\right) \tag{3}$$

allows us to write Hamilton's equations in commutator form. With Sudarshan[6] we define

$$x_2 = i\frac{\partial}{\partial p_1} \tag{4}$$

and

$$p_2 = -i\frac{\partial}{\partial x_1} \tag{5}$$

so that

$$[x_1, x_2] = [p_1, p_2] = 0\,, \tag{6}$$

$$[x_1, p_2] = [x_2, p_1] = i\,, \tag{7}$$

and $\hat{H}$ can now be taken to be the operator

$$\hat{H} = \frac{\partial H(x_1, p_1)}{\partial p_1}p_2 + \frac{\partial H(x_1, p_1)}{\partial x_1}x_2\,. \tag{8}$$

It is now apparent that we can take the above quantities and equations of motion to describe a quantum mechanical system with two degrees of freedom in the "coordinate" representation where the "coordinates" are (x_1, p_1) and the canonical momenta are $\Pi = (p_2, -x_2)$. As we will see below the linearity of $\hat{H}$ in the momenta is crucial for the maintenance of the classical character of x_1 and p_1, and for the observability of the phase-space trajectory. Since we choose to identify the physical observables with the commutative algebra of the coordinate operators, x_1 and p_1, we are led to impose the superselection condition that the momenta, Π, are unobservable. As a result the Hamiltonian and other generators of canonical transformations, which are all linear in the momenta, are also unobservable. However, in each case there is an associated dynamical quantity which is observable.

The required unobservability of the momenta restricts the form of the interaction between a classical-made-quantum system and an inherently quantum system to

$$H_{\rm int} = \Phi_1 x_2 + \Phi_2 p_2 + X\,, \tag{9}$$

where Φ_1, Φ_2, and X are functions of x_1, p_1, and the quantum system variables. The commutation relations of these functions are also constrained[6] by the superselection rule and the commutativity of the classical variables, x_1 and p_1, and their time derivatives. In the next section we will study the simple harmonic oscillator in order to exemplify the quantum-mechanical case described above and also for direct use in the field-theoretic generalizations of subsequent sections.

II. SIMPLE HARMONIC OSCILLATOR

In this section we discuss the embedding of a classical simple harmonic oscillator in a quantum system. We shall see that the space of states for the indefinite-metric classical-made-quantum system is far larger than the set of states of a classical harmonic oscillator. However, there is a subset of coherent states which may be placed in one-to-one correspondence with the classical harmonic-oscillator states. The classical-made-quantum oscillator is necessarily an indefinite-metric quantum theory for the simple physical reason that the classical bound states cannot have quantized energy levels. Indefinite-metric quantum theories normally have severe problems of physical interpretation. The present work raises the possibility of a partial resolution of some of these problems through a reinterpretation of an indefinite-metric quantum system as a system composed of a classical subsystem interacting with an essentially quantum subsystem of positive metric.

The classical simple harmonic oscillator of frequency ω has the Hamiltonian

$$\mathcal{H} = \frac{1}{2m}(p_1{}^2 + m^2\omega^2 x_1{}^2)\,, \tag{10}$$

and the motion is described by

$$x_1 = A\sin(\pi t + \delta)\,, \tag{11}$$

where A and δ are constants. To embed this classical system in a quantum-mechanical system we introduce the variables x_2 and p_2, and, using Eq. (8), obtain the quantum Hamiltonian

$$\hat{H} = \frac{1}{m}p_1 p_2 + m\omega^2 x_1 x_2\,. \tag{12}$$

We eliminate constants by defining (for $i = 1, 2$)

$$x_i = \left(\frac{1}{m\omega}\right)^{1/2} Q_i\,, \tag{13}$$

$$p_i = (m\omega)^{1/2} P_i\,, \tag{14}$$

and

$$\hat{H} = H\omega \tag{15}$$

so that

$$H = P_1 P_2 + Q_1 Q_2\,. \tag{16}$$

The raising and lowering operators are defined by

$$a_j = \frac{1}{\sqrt{2}}(Q_j + iP_j)\,, \tag{17}$$

and

$$a_j^\dagger = \frac{1}{\sqrt{2}}(Q_j - iP_j) \tag{18}$$

for $j = 1, 2$. They have the commutation relations

$$[a_i, a_j] = [a_i^\dagger, a_j^\dagger] = 0\,, \tag{19}$$

$$[a_i, a_j^\dagger] = 1 - \delta_{ij} \tag{20}$$

for $i,j=1,2$. As a result H is seen to have the form

$$H=\tfrac{1}{2}(a_1a_2^\dagger+a_2a_1^\dagger+a_1^\dagger a_2+a_2^\dagger a_1)\,. \tag{21}$$

The number operators are defined by

$$N_1=a_2a_1^\dagger \tag{22}$$

and

$$N_2=a_2^\dagger a_1 \tag{23}$$

and are not Hermitian. However, their sum is Hermitian and we see that

$$H=N_1+N_2\,. \tag{24}$$

The number operators have the following commutation relations with the raising and lowering operators:

$$N_ia_j=a_j(N_i+\delta_{ij}-1) \tag{25}$$

and

$$N_ia_j^\dagger=a_j^\dagger(N_i-\delta_{ij}+1) \tag{26}$$

for $i,j=1,2$.

Up to this point we have maintained a symmetry of the dynamics under the exchange of the subscripts, $1\leftrightarrow 2$. Now we must break that symmetry by choosing a vacuum state which is an eigenstate of Q_1 and P_1 or alternately a_1 and $a_1^\dagger$. The commutativity of Q_1 and P_1 permit this. The observability of Q_1 and P_1 for all time requires it. So we define

$$a_1^\dagger|0\rangle=a_1|0\rangle=0\,. \tag{27}$$

As a result $a_2|0\rangle\neq 0$ and $a_2^\dagger|0\rangle\neq 0$. The eigenstates of the number operators are

$$|n_+,n_-\rangle=(a_2^\dagger)^{n_+}(a_2)^{n_-}|0,0\rangle \tag{28}$$

and satisfy

$$N_1|n_+,n_-\rangle=-n_-|n_+,n_-\rangle\,, \tag{29}$$

$$N_2|n_+,n_-\rangle=n_+|n_+,n_-\rangle\,, \tag{30}$$

so that

$$H|n_+,n_-\rangle=(n_+-n_-)|n_+,n_-\rangle\,. \tag{31}$$

The lack of a lower bound to the energy spectrum is in a sense a problem but a necessary one in that it leads to the possibility of bound states with a continuous energy spectrum—a requirement of a faithful representation of the classical oscillator states. There is a subset of coherent states which can be put in a one-to-one relation with the set of classical oscillator states. The defining property of that subset is that its elements are eigenstates of the operators a_1 and $a_1^\dagger$. If we expand an element of that subset in terms of the number eigenstates

$$|z\rangle=\sum_{n_+,n_-=0}^{\infty}f(z|n_+,n_-)|n_+,n_-\rangle \tag{32}$$

and use

$$a_1^\dagger|n_+,n_-\rangle=-n_-|n_+,n_--1\rangle\,, \tag{33}$$

$$a_1|n_+,n_-\rangle=n_+|n_+-1,n_-\rangle \tag{34}$$

to evaluate the eigenvalue equations

$$a_1|z\rangle=iz^*|z\rangle\,, \tag{35}$$

$$a_1^\dagger|z\rangle=-iz|z\rangle\,, \tag{36}$$

we find

$$f(z|n_+,n_-)=\frac{C(iz^*)^{n_+}(iz)^{n_-}}{n_+!\,n_-!}\,, \tag{37}$$

where C is a constant. As a result

$$|z\rangle=C\exp[i(za_2+z^*a_2^\dagger)]|0,0\rangle\,. \tag{38}$$

We shall call the $|z\rangle$ states coherent states because of their close formal resemblance to the coherent states used in the study of the classical limit of harmonic oscillators, and of quantum electrodynamics[11] (which were eigenstates of the lowering operator but not of the raising operator).

Since $[H,a_1]=-a_1$, and $[H,a_1^\dagger]=a_1^\dagger$, it is clear that the (x_1,p_1) phase-space trajectory is sharp on the set of coherent $|z\rangle$ states. The classical trajectory represented by the state $|z\rangle$ is easily seen to be

$$x_1=\left(\frac{2}{m\omega}\right)^{1/2}R\sin(\omega t+\delta) \tag{39}$$

and

$$p_1=(2m\omega)^{1/2}R\cos(\omega t+\delta)\,, \tag{40}$$

where $z=Re^{i\delta}$. The linearity of H in the "momenta", $\Pi=(p_2,-x_2)$, is crucial for the observability of the phase-space trajectory. In fact, the linearity of all generators of canonical transformations in the momenta is necessary if the canonical transformations are not to take states out of the subset of coherent states.

The superselection rule which follows from the unobservability of the momenta, Π, is best approached by a consideration of the momentum-and coordinate-space representations of the coherent states. In the coordinate-space representation we find that Eqs. (35) and (36) give

$$\left[\left(\frac{m\omega}{2}\right)^{1/2}x_1+i\left(\frac{1}{2m\omega}\right)^{1/2}p_1\right]\langle x_1p_1|z\rangle=iz^*\langle x_1p_1|z\rangle \tag{41}$$

and

$$\left[\left(\frac{m\omega}{2}\right)^{1/2}x_1-i\left(\frac{1}{2m\omega}\right)^{1/2}p_1\right]\langle x_1p_1|z\rangle=-iz\langle x_1p_1|z\rangle\,, \tag{42}$$

so that

$$\langle x_1 p_1 | z\rangle = \sqrt{2}\,\delta\left(x_1 - \left(\frac{2}{m\omega}\right)^{1/2}\mathrm{Im}z\right)$$
$$\times\,\delta(p_1 - (2m\omega)^{1/2}\mathrm{Re}z)\,. \qquad (43)$$

We have normalized $\langle x_1 p_1|z\rangle$ so that

$$\langle z'|z\rangle = \int_{-\infty}^{\infty} dx_1 dp_1 \langle z'|x_1 p_1\rangle\langle x_1 p_1|z\rangle$$
$$= \delta(\mathrm{Re}z - \mathrm{Re}z')\delta(\mathrm{Im}z - \mathrm{Im}z')\,. \qquad (44)$$

In momentum space Eqs. (35) and (36) lead to the differential equations

$$\left[\left(\frac{m\omega}{2}\right)^{1/2} i\frac{d}{dp_2} + \left(\frac{1}{2m\omega}\right)^{1/2}\frac{d}{dx_2}\right]\langle x_2 p_2|z\rangle = iz^*\langle x_2 p_2|z\rangle \qquad (45)$$

and

$$\left[\left(\frac{m\omega}{2}\right)^{1/2} i\frac{d}{dp_2} - \left(\frac{1}{2m\omega}\right)^{1/2}\frac{d}{dx_2}\right]\langle x_2 p_2|z\rangle = -iz\langle x_2 p_2|z\rangle\,. \qquad (46)$$

They are easily integrated to give

$$\langle x_2 p_2|z\rangle = \frac{1}{\sqrt{2\pi}}\exp\left[-ip_2\left(\frac{2}{m\omega}\right)^{1/2}\mathrm{Im}z + ix_2(2m\omega)^{1/2}\mathrm{Re}z\right] \qquad (47)$$

with the normalization condition

$$\langle z'|z\rangle = \int_{-\infty}^{\infty} dx_2 dp_2 \langle z'|x_2 p_2\rangle\langle x_2 p_2|z\rangle$$
$$= \delta(\mathrm{Re}z - \mathrm{Re}z')\delta(\mathrm{Im}z - \mathrm{Im}z')\,. \qquad (48)$$

The transformation function between the two representations is

$$\langle x_1 p_1|x_2 p_2\rangle = \frac{1}{2\pi}\exp(+ip_2 x_1 - ip_1 x_2)\,, \qquad (49)$$

so that

$$\langle x_1 p_1|z\rangle = \int_{-\infty}^{\infty} dx_2 dp_2 \langle x_1 p_1|x_2 p_2\rangle\langle x_2 p_2|z\rangle\,. \qquad (50)$$

Each coherent state, $|z\rangle$, is a superselection sector in itself. There is no measurable dynamical variable $F = F(a_1, a_1^\dagger)$ which connects different states:

$$\langle z'|F(a_1, a_1^\dagger)|z\rangle = F(iz^*, -iz)\delta^2(z - z')\,. \qquad (51)$$

This reflects the lack of a superposition principle in classical mechanics.

The operator formalism for coherent states is incomplete in that we have not defined an inner product. To remedy this deficiency we define the vacuum dual to $|0,0\rangle$ to satisfy

$$\langle 0,0|a_2 = \langle 0,0|a_2^\dagger = 0 \qquad (52)$$

with $\langle 0,0|0,0\rangle = 1$. The dual state corresponding to the physical state, z, we define to be

$$\langle z| = \langle 0,0|\delta(ia_1 + z^*)\delta(ia_1^\dagger - z)$$
$$= \langle 0,0|\int_{-\infty}^{\infty}\frac{d\alpha d\beta}{(2\pi)^2}\exp[i\alpha(\mathrm{Im}z - 2^{-1/2}Q_1) + i\beta(\mathrm{Re}z - 2^{-1/2}P_1)] \qquad (53)$$

so that Eqs. (48) and (51) follow if we choose $C = 1$.

Sometimes the dynamical state of a classical system is incompletely known and one only has a set of probabilities that the system is at a particular phase-space point at $t = 0$. If we let $P(z)$ be the probability that the system is at a phase-space point corresponding to z (as defined above), then using the properties

$$P(z) \geq 0, \quad \int d^2z\, P(z) = 1 \qquad (54)$$

one sees that a density operator

$$\rho\delta^2(0) = \int d^2z\, |z\rangle P(z)\langle z| \qquad (55)$$

may be defined which satisfies

$$\mathrm{Tr}\rho = 1 \qquad (56)$$

and

$$\langle z'|\rho|z'\rangle \equiv \lim_{z''\to z'}\langle z''|\rho|z'\rangle = P(z')\,. \qquad (57)$$

The mean value of an observable $A = A(a_1, a_1^\dagger)$ is given by

$$\langle A\rangle = \mathrm{Tr}\rho A = \int d^2z\, A(iz^*, -iz)P(z)\,, \qquad (58)$$

and one can develop a formalism similar to the density-matrix formalism of quantum mechanics.

We now turn to a closer investigation of the relation of the pseudoquantum mechanics discussed above and true quantum-mechanical systems. We shall be particularly interested in the relation of the coherent states described above and the coherent states of a quantum-mechanical harmonic oscillator—to which they bear such a remarkable resemblance. We shall see that the pseudoquantum oscillator system is equivalent to an indefinite-metric quantum system composed of a harmonic oscillator (thus the connection to the coherent-state quantum oscillator formalism) and an "inverted" oscillator to be described below.

Let us define the following rotated raising and lowering operators in terms of the operators defined in Eqs. (17) and (18):

$$b_1 = a_1\cos\theta + a_2\sin\theta\,, \qquad (59)$$

$$b_2 = -a_1\sin\theta + a_2\cos\theta\,. \qquad (60)$$

Their commutation relations are

$$[b_1, b_1^\dagger] = \sin(2\theta)\,, \tag{61}$$

$$[b_2, b_2^\dagger] = -\sin(2\theta)\,, \tag{62}$$

$$[b_2, b_1^\dagger] = [b_1, b_2^\dagger] = \cos(2\theta) \tag{63}$$

with all other commutators equal to zero. The Hamiltonian of Eq. (21) becomes

$$H = \tfrac{1}{2}(\{b_1, b_1^\dagger\} - \{b_2, b_2^\dagger\})\sin(2\theta) + \tfrac{1}{2}(\{a_1, a_2^\dagger\} + \{a_2, a_1^\dagger\})\cos(2\theta)\,, \tag{64}$$

where $\{u, v\} = uv + vu$.

Now θ is an arbitrary angle and it is obvious that choosing $\theta = 0$ gives the commutation relations and Hamiltonian studied above. However, the choice $\theta = \pi/4$ results in a new form for H and the commutation relations, which can be interpreted as a harmonic oscillator (the b_1 and $b_1^\dagger$ sector) and an "inverted" harmonic oscillator (the b_2 and $b_2^\dagger$ sector) where the commutator and b_2 terms in the Hamiltonian have the wrong sign. The commutativity of the oscillator raising and lowering operators with the inverted oscillator raising and lowering operators leads to a simple factorization of the coherent states which lays bare the basic of the close similarity of form for our coherent states and the coherent states of a quantum oscillator[10]:

$$|z\rangle = \frac{1}{\sqrt{2\pi}} \exp\left[\frac{i}{\sqrt{2}}(z b_1 + z^* b_1^\dagger)\right] \times \exp\left[\frac{i}{\sqrt{2}}(z b_2 + z^* b_2^\dagger)\right] |0, 0\rangle\,, \tag{65}$$

while the coherent state of Ref. 11 has the form

$$|\alpha\rangle = \exp(\alpha b^\dagger - \alpha^* b)|0\rangle\,, \tag{66}$$

where α is a complex numer and $[b, b^\dagger] = 1$. It should be remembered that our choice of vacuum state such that $a_1|0, 0\rangle = a_1^\dagger|0, 0\rangle = 0$ obviates a simple direct relationship.

Since we have uncovered an interesting relation between a classical-made-quantum system and a "quantum" system of indefinite metric the possibility of reinterpreting indefinite-metric quantum systems as systems containing classical subsystems naturally arises.

III. EMBEDDING OF CLASSICAL FIELDS

In this section we shall discuss the embedding of a classical field theory in a quantum field theory. We shall study the embedding in detail for a scalar field and then describe the features of a classical-made-quantum electrodynamics which we shall call pseudoquantum electrodynamics for the sake of brevity.

Consider a classical field, $\phi_1(x)$, with canonically conjugate momentum, $\pi_1(x)$, and Hamiltonian equations of motion

$$\frac{d}{dt}\phi_1(x) = \frac{\delta\hat{H}}{\delta\pi_1(x)}\,, \tag{67}$$

$$\frac{d}{dt}\pi_1(x) = \frac{-\delta H}{\delta\phi_1(x)}\,, \tag{68}$$

where $\hat{H}$ is the Hamiltonian. We wish to define a "quantum" Hamiltonian, H, which allows us to rewrite Eqs. (67) and (68) in commutator form:

$$\frac{d}{dt}\phi_1(x) = i[H, \phi_1(x)]\,, \tag{69}$$

$$\frac{d}{dt}\pi_1(x) = i[H, \pi_1(x)]\,. \tag{70}$$

Equations (69) and (70) are satisfied if

$$H = \int d^3x \left[\frac{\delta H}{\delta\pi_1(x)} \frac{1}{i} \frac{\delta}{\delta\phi_1(x)} - \frac{\delta H}{\delta\phi_1(x)} \frac{1}{i} \frac{\delta}{\delta\pi_1(x)}\right]. \tag{71}$$

We now formally define

$$\phi_2(x) = i\frac{\delta}{\delta\pi_1(x)} \tag{72}$$

and

$$\pi_2(x) = -i\frac{\delta}{\delta\phi_1(x)}\,, \tag{73}$$

so that

$$H = \int d^3x \left[\frac{\delta\hat{H}}{\delta\pi_1(x)}\pi_2(x) + \frac{\delta\hat{H}}{\delta\phi_1(x)}\phi_2(x)\right]. \tag{74}$$

The fields satisfy the equal-time commutation relations

$$[\phi_i(x), \pi_j(y)] = i(1 - \delta_{ij})\delta^3(\vec{x} - \vec{y})\,, \tag{75}$$

$$[\phi_i(x), \phi_j(y)] = 0\,, \tag{76}$$

$$[\pi_i(x), \pi_j(y)] = 0\,, \tag{77}$$

where δ_{ij} is the Kronecker δ.

We note that the linearity of H in ϕ_2 and π_2 is necessary to maintain the classical character of ϕ_1 and π_1. This is best seen by an examination of Eqs. (69) and (70) and the corresponding Hamiltonian equations for ϕ_2 and π_2. (Other generators of canonical transformations are also linear in π_2 and ϕ_2.)

$\phi_2(x)$ and $\pi_2(x)$ will not be observables on the set of physical states, so that $\phi_1(x)$ and $\pi_1(x)$ will both be sharp on the set of physical states and satisfy superselection rules.

If we wish to couple the classical field to a truly quantum system and maintain the classical nature of the field then certain restrictions exist on the form of the total Hamiltonian H_{tot} and on the commutation relations of the various terms occurring in it. First, the coupling must satisfy the requirement that H_{tot} is linear in $\phi_2(x)$ and $\pi_2(x)$. If we denote the quantum fields by ψ and write the general form of the Hamiltonian as

$$H_{tot} = H + H_Q(\psi) + H_{int}, \tag{78}$$

where H is given by Eq. (74), $H_Q(\psi)$ depends only on the quantum fields, ψ, and

$$H_{int} = \int d^3x[\tilde{A}(\phi_1, \pi_1, \psi)\phi_2(x) + \tilde{B}(\phi_1, \pi_1, \psi)\pi_2(x) + \tilde{C}(\phi_1, \pi_1, \psi)], \tag{79}$$

then we can rearrange the Hamiltonian so that

$$H_{tot} = \int d^3x[A(\phi_1, \pi_1, \psi)\phi_2(x) + B(\phi_1, \pi_1, \psi)\pi_2(x) + C(\phi_1, \pi_1, \psi)], \tag{80}$$

where

$$A = \frac{\delta \hat{H}}{\delta \phi_1(x)} + \tilde{A}, \tag{81}$$

$$B = \frac{\delta \hat{H}}{\delta \pi_1(x)} + \tilde{B}, \tag{82}$$

and

$$C = \tilde{C} + \mathcal{H}_Q \tag{83}$$

with $H_Q = \int d^3x \mathcal{H}_Q$. An examination of the equations of motion of $\phi_1(x)$, $\pi_1(x)$, and ψ,

$$\frac{d}{dt}\phi_1 = B(\phi_1, \pi_1, \psi), \tag{84}$$

$$\frac{d}{dt}\pi_1 = A(\phi_1, \pi_1, \psi), \tag{85}$$

$$\frac{d}{dt}\psi = i[H_{tot}, \psi], \tag{86}$$

and the second time derivatives of ϕ_1 and π_1, such as

$$\frac{d^2}{dt^2}\phi_1(x) = i[H, B] = \int d^3y\left(-A\frac{\delta B}{\delta \pi_1(y)} + B\frac{\delta B}{\delta \phi_1(y)} + i\phi_2(y)[A, B] + i\pi_2(y)[B(y), B(x)] + i[C, B]\right), \tag{87}$$

leads us to require the equal-time commutation relations

$$[A(x), A(y)] = [A(x), B(y)] = [B(x), B(y)] = 0, \tag{88}$$

where $A(x) = A(\phi_1(x), \pi_1(x), \psi(x))$, etc., so that $\phi_1(x)$ and $\pi_1(x)$ are independent of ϕ_2 and π_2 and hence observable for all time. An examination of higher time derivatives of ϕ_1 and π_1 lead to further restrictions on the equal-time commutation relations of A, B, and C. Examples are

$$[A, [C, B]] = 0, \tag{89}$$

$$[B, [C, B]] = 0, \tag{90}$$

$$[A, [C, [C, [C, B]]]] = 0, \tag{91}$$

etc. A sufficient condition for satisfying all relations of this class consists of having equal-time commutation relations with the form

$$[A, C] = F_1(A, B, \phi_1, \pi_1) \tag{92}$$

and

$$[B, C] = F_2(A, B, \phi_1, \pi_1). \tag{93}$$

Finally, we note that another obvious requirement [cf. Eqs. (84) and (85)] for the observability of ϕ_1 and π_1 is that A and B depend only on an (equal-time) commutative subset of the quantum field variables, ψ.

The above restrictions on the equal-time commutation relations have a direct interpretation in terms of Feynman diagrams for quantum corrections to the classical field behavior. For example, consider the interaction of the classical field sector with a scalar quantum field, ψ, expressed in the interaction

$$H_{int} = g\phi_2(x)\psi^2(x). \tag{94}$$

If $H_Q(\psi)$ is the conventional free Klein-Gordon Hamiltonian, then we find that Eq. (92) is not satisfied so that the Green's function for the classical ϕ_1 field receives quantum corrections from vacuum polarization loops of ψ particles and thus loses its classical character.

We now define a Lagrangian appropriate to our pseudoquantum field theory and then verify the reasonableness of our definition, and the pseudoquantization procedure described above, by studying the equivalent path-integral formulation. The Lagrangian corresponding to the pseudoquantum Hamiltonian, H, is

$$L = \int d^3x(\pi_1\dot{\phi}_2 + \pi_2\dot{\phi}_1) - H, \tag{95}$$

where $L = L(\phi_1, \dot{\phi}_1, \phi_2, \dot{\phi}_2)$ and

$$\pi_1 = \frac{\delta L}{\delta \dot{\phi}_2}, \tag{96}$$

$$\pi_2 = \frac{\delta L}{\delta \dot{\phi}_1}. \tag{97}$$

The vacuum-vacuum transition amplitude for the field theory corresponding to the H_{tot} of Eq. (78) will be shown to be

$$W = \int \prod_x d\phi_1(x) d\phi_2(x) d\pi_1(x) d\pi_2(x) d\psi(x) \exp(iS), \tag{98}$$

where $S = \int dt\, L_{tot}$ up to external source terms. We begin by considering the vacuum-vacuum transition amplitude corresponding to H_Q,

$$W_Q = \int \prod_x d\psi(x) \exp(iS_Q), \tag{99}$$

where ϕ_1 has the character of an external source.

We can now introduce the classical behavior of the ϕ_1 field through functional δ functions

$$\int \prod_x d\psi(x) d\phi_1(x) d\pi_1(x) \delta(B(\phi_1, \pi_1, \psi) - \dot{\phi}_1) \times \delta(A(\phi_1, \pi_1, \psi) + \dot{\pi}_1) e^{iS_Q}, \tag{100}$$

which can be put in the form

$$\int \prod_x d\phi_1(x) d\pi_1(x) d\phi_2(x) d\pi_2(x) \times \exp\left\{ i \int d^4x [(\dot{\phi}_1 - B)\pi_2 - (\dot{\pi}_1 + A)\phi_2] + iS_Q \right\}. \tag{101}$$

After performing a partial integration on the $\dot{\pi}_1\phi_2$ term and discarding a surface term we see that the definition of L in Eq. (95) is correct and that the vacuum-vacuum transition amplitude is indeed given by Eq. (98).

The restrictions on the commutation relations of the various terms in the H_{tot} [expressed in Eqs. (88)-(93)] translate into the requirement that the "quantum completion"[11] of the ϕ_2 field does not take place, i.e., that all N-point functions of the ϕ_2 field are zero:

$$\frac{\delta^n W}{\delta J_2(x_1) \delta J_2(x_2) \cdots \delta J_2(x_n)} = 0, \tag{102}$$

where J_2 is an external source coupled to ϕ_2.

We now discuss the embedding of a free classical Klein-Gordon field in a quantum field theory. The Lagrangian density is

$$\mathcal{L} = \frac{\partial \phi_1}{\partial x^\mu} \frac{\partial \phi_2}{\partial x_\mu} - m^2 \phi_1 \phi_2, \tag{103}$$

from which one obtains the Euler-Lagrange equations (for $i = 1, 2$)

$$(\Box + m^2)\phi_i(x) = 0. \tag{104}$$

The canonical momenta are (note that π_2 is conjugate to ϕ_1, etc.)

$$\Pi_i = \dot{\phi}_i \tag{105}$$

for $i = 1, 2$ with the equal-time commutation relations given by Eqs. (75)-(77). We expand the fields in Fourier integrals:

$$\phi_1(\vec{x}, t) = \int d^3k [a_1(k) f_k(x) + a_1^\dagger f_k^*(x)] \tag{106}$$

and

$$\phi_2(\vec{x}, t) = \int d^3k [a_2(k) f_k(x) + a_2^\dagger(k) f_k^*(x)], \tag{107}$$

where

$$f_k(x) = (2\pi)^{-3/2} (2\omega_k)^{-1/2} e^{-ik\cdot x} \tag{108}$$

with $\omega_k = (\vec{k}^2 + m^2)^{1/2}$. The Fourier component operators satisfy the commutation relations

$$[a_i(k), a_j^\dagger(k')] = (1 - \delta_{ij}) \delta^3(\vec{k} - \vec{k}') \tag{109}$$

and

$$[a_i(k), a_j(k')] = [a_i^\dagger(k), a_j^\dagger(k')] = 0 \tag{110}$$

for $i, j = 1, 2$.

In terms of the Fourier coefficients

$$H = \int d^3x (\dot{\phi}_1 \dot{\phi}_2 + \vec{\nabla}\phi_1 \cdot \vec{\nabla}\phi_2 + m^2 \phi_1 \phi_2) \tag{111}$$

becomes

$$H = \int d^3k\, \omega_k [\{a_1(k), a_2^\dagger(k)\} + \{a_2(k), a_1^\dagger(k)\}]. \tag{112}$$

The analogy between the mode amplitudes of the fields and the raising and lowering operators of the simple harmonic oscillator has been previously remarked. We can therefore use the considerations of Sec. II to establish the spectrum of physical states. The defining properties of a physical state are that $\phi_1(x)$ and $\pi_1(x)$ are sharp on it for all time:

$$\phi_1(x)|\Phi, \Pi\rangle = \Phi(x)|\Phi, \Pi\rangle \tag{113}$$

and

$$\pi_1(x)|\Phi, \Pi\rangle = \Pi(x)|\Phi, \Pi\rangle, \tag{114}$$

where $\Phi(x)$ and $\Pi(x)$ are c-number functions of x:

$$\Phi(x) = \int d^3k [\alpha(k) f_k(x) + \alpha^*(k) f_k^*(x)] \tag{115}$$

and

$$\Pi(x) = -i \int d^3k\, \omega_k [\alpha(k) f_k(x) - \alpha^*(k) f_k^*(x)] \tag{116}$$

with $\alpha(k)$ a c-number function of k.

As a result we are led to define a set of physical states, $|\alpha\rangle$, which are in one-to-one correspondence with the classical solutions of the Klein-Gordon equation and satisfy

$$a_1(k)|\alpha\rangle = \alpha(k)|\alpha\rangle, \tag{117}$$

$$a_1^\dagger(k)|\alpha\rangle = \alpha^*(k)|\alpha\rangle. \tag{118}$$

In analogy with the states of the simple harmonic oscillator (Sec. II) we further define

$$|\alpha\rangle = C\exp\left\{\int d^3k'[\alpha(k')a_2^\dagger(k') - \alpha^*(k')a_2(k')]\right\}|0\rangle, \tag{119}$$

where the vacuum state, $|0\rangle$, satisfies

$$a_1(k)|0\rangle = a_1^\dagger(k)|0\rangle = 0. \tag{120}$$

The physical states, $|\alpha\rangle$, lie in a space which is the infinite tensor product of single-mode spaces. While ϕ_1 and π_1 are sharp for all time on the subset of physical states, we see that ϕ_2 and π_2 are not and, in fact, when applied to a physical state map it into an unphysical state. The superselection rules are embodied in

$$\langle\alpha'|\mathcal{O}|\alpha\rangle = \mathcal{O}_\alpha\delta^2(\alpha - \alpha'), \tag{121}$$

where $\mathcal{O}$ is the operator corresponding to any observable, $\mathcal{O}_\alpha$ is its eigenvalue for the state $|\alpha\rangle$, and $\delta^2(\alpha - \alpha')$ is a functional δ function in the real and imaginary parts of $\alpha - \alpha'$. The functional δ functions have their origin in the definition of the dual set of physical states. We define the dual vacuum state $\langle 0|$ by

$$\langle 0|a_2(k) = 0 \tag{122a}$$

and

$$\langle 0|a_2^\dagger(k) = 0 \tag{122b}$$

for all k with $\langle 0|0\rangle = 1$. The dual state corresponding to $\alpha(k)$ we define by

$$\langle\alpha| = \langle 0|\prod_k \delta(\alpha(k) - a_1(k))\delta(\alpha^*(k) - a_1^\dagger(k))$$

$$\equiv \langle 0|\delta(\alpha - a_1)\delta(\alpha^* - a_1^\dagger), \tag{123}$$

so that

$$\langle\alpha'|\alpha\rangle = \delta^2(\alpha' - \alpha) \tag{124}$$

if $C = 1$.

We have now established a procedure for embedding a classical field in a quantum field theory. Given a Lagrangian, L, for a classical field theory describing a field $\phi_1(x)$, the Lagrangian density for the pseudoquantum field theory, $\mathcal{L}_{PQ}$ is

$$\mathcal{L}_{PQ}(\phi_1, \dot\phi_1, \phi_2, \dot\phi_2) = \frac{\delta L}{\delta\phi_1(x)}\phi_2(x) + \frac{\delta L}{\delta\dot\phi_1(x)}\pi_2(x) \tag{125}$$

up to a divergence with

$$\pi_2(x) = \frac{\delta}{\delta\dot\phi_1(x)}\int d^3x\,\mathcal{L}_{PQ}. \tag{126}$$

In the case of a classical electromagnetic field interacting with a quantum electron field, one pseudoquantum model, which describes some electromagnetic processes, has the Lagrangian

$$\mathcal{L} = -\tfrac{1}{2}F^1_{\mu\nu}F^2_{\mu\nu} + \bar\psi(i\nabla\!\!\!/ - eA\!\!\!/_1 - m_0)\psi, \tag{127}$$

where $A^1_\mu(x)$ is the classical electromagnetic field, ψ is the electron field, $A^2_\mu(x)$ is the unobservable auxiliary field, and $F^i_{\mu\nu} = \partial_\nu A^i_\mu - \partial_\nu A^i_\mu$ for $i = 1, 2$. Although our interpretation of the free electromagnetic part of the Lagrangian, $-\frac{1}{2}F^1_{\mu\nu}F^2_{\mu\nu}$, is new, the actual form of this term appeared some time ago in a generalization of electrodynamics by Mie,[12] and was recently used in an Abelian prototype model for quark confinement.[8] The equations of motion are

$$\partial^\mu F^1_{\mu\nu} = 0, \tag{128}$$

$$\partial^\mu F^2_{\mu\nu} + eJ_\nu = 0, \tag{129}$$

and

$$(i\nabla\!\!\!/ - eA\!\!\!/^1 - m)\psi = 0. \tag{130}$$

The canonical momentum which is conjugate to A^1_μ is

$$\Pi^2_\mu = F^2_{0\mu} \tag{131}$$

and that conjugate to A^2_μ is

$$\Pi^1_\mu = F^1_{0\mu}. \tag{132}$$

We take A^1_μ and Π^1_μ to be classical fields which are observable for all time. A^2_μ and Π^2_μ are not observable. Note that $\mathcal{L}$ is invariant under the independent gauge transformations

$$A^1_\mu \to A^1_\mu + \partial_\mu\Lambda^1(x) \tag{133}$$

and

$$A^2_\mu \to A^2_\mu + \partial_\mu\Lambda^2(x). \tag{134}$$

Since $\Pi^1_0 = \Pi^2_0 = 0$, it is apparent that A^1_0 and A^2_0 are c numbers. If we chose the Coulomb gauge for A^1_μ,

$$\vec\nabla\cdot\vec{A}^1 = 0, \tag{135}$$

and for A^2_μ,

$$\vec\nabla\cdot\vec{A}^2 = 0, \tag{136}$$

then we can establish the equal-time commutation relations

$$[\Pi^a_i(\vec{x},t),A^b_j(\vec{y},t)]=i(1-\delta_{ab})\times\int\frac{d^3k}{(2\pi)^3}e^{i\vec{k}\cdot(\vec{x}-\vec{y})}\left(\delta_{ij}-\frac{k_ik_j}{|\vec{k}|^2}\right)=i(1-\delta_{ab})\delta^{tr}_{ij}(\vec{x}-\vec{y}) \quad (137)$$

for $a,b=1,2$ and $i,j=1,2,3$.

This pseudoquantum field theory describes the dynamics of quantum electron fields interacting with a free, classical electromagnetic field. A typical perturbation theory matrix element would have the form

$$\langle\mathfrak{A}',0|T(\bar{\psi}(x)J^{\mu_1}(x_1)A^1_{\mu_1}(x_1)J^{\mu_2}(x_2)A^1_{\mu_2}(x_2)\cdots J^{\mu_n}(x_n)A^1_{\mu_n}(x_n)\psi(y))|\mathfrak{A},0\rangle, \quad (138)$$

where $|\mathfrak{A},0\rangle$ is the tensor product of an electron vacuum state and an electromagnetic state corresponding to the classical field $\mathfrak{A}_\mu(z)$. Because $A^1_\mu(x)$ is sharp on this state, the matrix element becomes

$$\langle 0|T(\bar{\psi}(x)J^{\mu_1}(x_1)\cdots J^{\mu_n}(x_n)\psi(y))|0\rangle\mathfrak{A}_{\mu_1}(x_1)\mathfrak{A}_{\mu_2}(x_2)\cdots\mathfrak{A}_{\mu_n}(x_n) \quad (139)$$

modulo a functional δ function in $\mathfrak{A}'-\mathfrak{A}$. Thus this model is equivalent to a quantized electron field interacting with an external electromagnetic field.

Another possibility for a model electrodynamics is realized by letting the interaction term in Eq. (127) above be replaced with

$$L_{\rm int}=-e\bar{\psi}A_2\psi. \quad (140)$$

Because the equivalent of the equal-time commutation relation, Eq. (92), is not true in this model, the A^1_μ field loses its purely classical character due to quantum corrections. However, this model may be of value for the study of the modification of the A^1_μ field resulting from the emission of many soft photons by a current.

Since vacuum polarization effects modify the electromagnetic field in this case we define in-field eigenstates (in the transverse gauge) by

$$\vec{A}^1_{\rm in}|\mathfrak{A}\rangle_{\rm in}=\vec{\mathfrak{A}}_{\rm in}|\mathfrak{A}\rangle_{\rm in}, \quad (141)$$

where

$$|\mathfrak{A}\rangle_{\rm in}=\exp\left[\int d^3k\sum_{\lambda=1}^{2}(\alpha(k,\lambda)a_2^\dagger(k,\lambda)-\alpha^*(k,\lambda)a_2(k,\lambda))\right]|0\rangle \quad (142)$$

and

$$\vec{\mathfrak{A}}_{\rm in}=\int d^3k\sum_{\lambda=1}^{2}\vec{\epsilon}(k,\lambda)[\alpha(k,\lambda)f_k(x)+\alpha^*(k,\lambda)f^*_k(x)] \quad (143)$$

with

$$\vec{A}^i_{\rm in}=\int d^3k\sum_{\lambda=1}^{2}\vec{\epsilon}(k,\lambda)[a_i(k,\lambda)f_k(x)+a_i^\dagger(k,\lambda)f^*_k(x)] \quad (144)$$

for $i=1,2$. The vacuum state is defined by

$$a_1(k,\lambda)|0\rangle=a_1^\dagger(k,\lambda)|0\rangle=0$$

for all k,λ. The interacting field, $\vec{A}^1$, is apparently not sharp on $|\mathfrak{A}\rangle_{\rm in}$ but is sharp on

$$|\mathfrak{A}\rangle=U^{-1}(t,-\infty)|\mathfrak{A}\rangle_{\rm in}, \quad (145)$$

where

$$U(t,-\infty)=T\left(\exp\left[-i\int_{-\infty}^{t}d^4x\,H_{\rm int}(A^2_{\rm in},\psi_{\rm in})\right]\right) \quad (146)$$

because

$$\vec{A}^1(\vec{x},t)=U^{-1}(t,-\infty)\vec{A}^1_{\rm in}(\vec{x},t)U(t,-\infty). \quad (147)$$

With these preliminaries completed, the study of physical processes within the framework of these models is now possible, although we shall not pursue it in this report.

Before turning to a discussion of non-Abelian gauge field theories, it is worth noting that the choice of vacuum state we have made necessitates a redefinition of normal-ordering. By normal-ordering a Lagrangian term we shall mean that the observable fields (to which we have consistently appended the superscript or subscript one) are to be placed to the right, and unobservable fields, labeled by two, are to be placed to the left. Thus Wick's theorem (with our definition of normal-ordering) becomes in the case of two fields

$$\begin{aligned}T(\phi_{1\,\rm in}(x_1)\phi_{2\,\rm in}(x_2))&=:\phi_{1\,\rm in}(x_1)\phi_{2\,\rm in}(x_2):\\&\quad+\langle 0|T(\phi_{1\,\rm in}(x_1)\phi_{2\,\rm in}(x_2))|0\rangle\\&=\phi_{2\,\rm in}(x_2)\phi_{1\,\rm in}(x_1)\\&\quad+\theta(x_{10}-x_{20})[\phi_{1\,\rm in}(x_1),\phi_{2\,\rm in}(x_2)].\end{aligned} \quad (148)$$

Note that the Green's function

$$G(x_1,x_2)=\langle 0|T(\phi_{1\,\rm in}(x_1)\phi_{2\,\rm in}(x_2))|0\rangle \quad (149)$$

is necessarily retarded. From this we can conclude that the models of electrodynamics, which we have considered, naturally embody the observed retarded nature of classical electrodynamics. Another way of stating this result is: If classical electrodynamics is to have a pseudoquantum formulation, its Green's functions are necessarily retarded. The origin of the asymmetry is the definition of the vacuums (which is equivalent to a specification of boundary conditions). Just as in classical electrodynamics retarded propagation is implemented by a choice of boundary conditions which do not require a commitment to any specific cosmological model.

Finally we would like to note that the Lagrangian obtained from adding L_{int} of Eq. (140) to the Lagrangian of Eq. (127) is equivalent to the usual Lagrangian of electrodynamics plus a term describing a massless Abelian gauge field with the wrong sign. (This is seen by defining new fields equal to the sum and difference of A^1_μ and A^2_μ.) This field theory may be quantized following the procedure we have outlined. A^1_μ loses its classical character due to quantum corrections.

IV. NON-ABELIAN GAUGE THEORIES

In this section we shall describe the procedure for embedding a classical non-Abelian Yang-Mills field in a quantum field theory. Then we will discuss a vierbein formulation of quantum gravity which could have been interpreted as a pseudoquantum field theory for a classical metric field if it were not for one term in the Lagrangian which makes it a truly quantum field theory. Nevertheless we suggest a new canonical quantization procedure based on our pseudoquantum approach.

Consider a classical Yang-Mills field, $A^1_\mu = \underline{A}^1_\mu \cdot \underline{T}$, where the jth component of $\underline{T}$ is a matrix representing a generator of a non-Abelian group G in the defining representation with commutation relations

$$[T_j, T_k] = i t_{jkl} T_l. \tag{150}$$

We can define a pseudoquantum field theory, wherein the classical character of A^1_μ is maintained, which has the Lagrangian density

$$\begin{aligned}\mathcal{L} = {} & \tfrac{1}{2}\underline{F}^1_{\mu\nu}\cdot\underline{F}^{2\mu\nu} - \tfrac{1}{2}\underline{F}^{2\mu\nu}\cdot(\partial_\mu \underline{A}^1_\nu - \partial_\nu \underline{A}^1_\mu + g\underline{A}^1_\mu \times \underline{A}^1_\nu) \\ & - \tfrac{1}{2}\underline{F}^{1\mu\nu}\cdot(\partial_\mu \underline{A}^2_\nu - \partial_\nu \underline{A}^2_\mu + g\underline{A}^1_\mu \times \underline{A}^2_\nu - g\underline{A}^1_\nu \times \underline{A}^2_\mu) \\ & + \bar{\psi}(i\nabla\!\!\!/ + g A\!\!\!/^1 - m)\psi,\end{aligned} \tag{151}$$

where ψ is a fermion field. The theory is invariant under the local gauge transformation, $S \in G$,

$$\psi' = S^{-1}\psi, \tag{152}$$

$$A^{1\prime}_\mu = S^{-1}A^1_\mu S + \frac{i}{g}S^{-1}\partial_\mu S, \tag{153}$$

$$F^{1\prime}_{\mu\nu} = S^{-1}F^1_{\mu\nu}S, \tag{154}$$

$$A^{2\prime}_\mu = S^{-1}A^2_\mu S, \tag{155}$$

$$F^{2\prime}_{\mu\nu} = S^{-1}F^2_{\mu\nu}S. \tag{156}$$

Except for one important term this Lagrangian with its attendant gauge invariance properties has been suggested as a possible model for the quark-confining strong interaction.[8] Since the omitted term has a masslike character $\Lambda^2\underline{A}^2_\mu \cdot \underline{A}^{2\mu}$, where Λ has the dimensions of a mass, it is clear that the strong-interaction model's ultraviolet behavior approaches that of the present pseudoquantum theory if the same quantization procedure is followed in both cases. We shall discuss this question further in the next section and show that the *ad hoc* procedure followed in Ref. 8 leads to the same result as the quantization procedure developed in this report.

The Euler-Lagrange equations of motion which are obtained from $\mathcal{L}$ in the canonical manner are

$$\underline{F}^1_{\mu\nu} = \partial_\mu \underline{A}^1_\nu - \partial_\nu \underline{A}^1_\mu + g\,\underline{A}^1_\mu \times \underline{A}^1_\nu, \tag{157}$$

$$\underline{F}^2_{\mu\nu} = \partial_\mu \underline{A}^2_\nu - \partial_\nu \underline{A}^2_\mu + g\underline{A}^1_\mu \times \underline{A}^2_\nu - g\underline{A}^1_\nu \times \underline{A}^2_\mu, \tag{158}$$

$$(\partial_\mu + g\,\underline{A}^1_\mu \times)\underline{F}^{1\mu\nu} = 0, \tag{159}$$

$$(\partial_\mu + g\underline{A}^1_\mu \times)\underline{F}^{2\mu\nu} + g\underline{A}^2_\mu \times \underline{F}^{1\mu\nu} + g\underline{J}^\nu = 0, \tag{160}$$

$$(i\nabla\!\!\!/ + g A\!\!\!/^1 - m)\psi = 0, \tag{161}$$

with the conservation law

$$(\partial_\nu + g\underline{A}^1_\nu \times)\underline{J}^\nu = 0. \tag{162}$$

The canonical momentum which is conjugate to $\underline{A}^1_j$ is

$$\underline{\Pi}^2_j = \underline{F}^2_{0j} \tag{163}$$

and the canonical momentum conjugate to $\underline{A}^2_j$ is

$$\underline{\Pi}^1_j = \underline{F}^1_{0j} \tag{164}$$

for $j = 1, 2, 3$. The canonical momentum corresponding to the fields $\underline{A}^i_0$ is zero for $i = 1, 2$. The existence of equations of constraint among the Euler-Lagrange equations implies that not all field components are independent, so that we must isolate the independent components prior to defining the canonical equal-time commutation relations.

Following Ref. 8 we choose to work in the Coulomb gauge, $\nabla_i \underline{A}^1_i = 0$, and define the field variables

$$\underline{A}^2_i = \underline{A}^{2T}_i + \underline{A}^{2L}_i, \tag{165}$$

$$\underline{\Pi}^a_i = \underline{\Pi}^{aT}_i + \underline{\Pi}^{aL}_i, \tag{166}$$

where

$$\nabla_i \cdot \underline{A}^{2T}_i = \nabla_i \cdot \underline{\Pi}^{aT}_i = 0 \tag{167}$$

and $a = 1, 2$. Then the nonzero equal-time commutation relations are

$$[\Pi^{aT}_{ip}(x), A^{bT}_{jq}(y)] = i\delta_{pq}(1 - \delta_{ab})\delta^{tr}_{ij}(\vec{x} - \vec{y}), \tag{168}$$

where p and q are internal-symmetry indices, $a, b = 1, 2$, and $i, j = 1, 2, 3$.

While the classical character of A^i_μ can be maintained with our choice of $\mathcal{L}$, this theory has features due to its non-Abelian nature which make it less trivial and therefore more interesting than the corresponding Abelian theory discussed in the last section. If we follow a procedure similar to that in the Abelian case [Eq. (127)] and introduce a set of states appropriate to the quadratic part of the Lagrangian, then the cubic and quartic Yang-Mills terms in the interaction part of the Lagrangian will act to transform $A^i_{\text{in}\,\mu}$ eigenstates into eigenstates of the interacting field A^i_μ. This is, of course, necessary for the classical Yang-Mills equations of motion to be satisfied. Our formalism, thus, offers a perturbative method for calculating solutions of the classical Yang-Mills equations. In addition, it gives an interesting interpretation to the short-distance behavior of the quark-confining field theory of Ref. 8. At short distances the gluon field A^i_μ effectively decouples from the quark sector and becomes, in effect, a free field. This type of short-distance behavior is certainly not at odds with the seemingly simple behavior observed in hadron processes at high energy. Therefore, it is possible that pseudoquantum field theory may be relevant to the short-distance behavior of hadron interaction. Certainly, it is interesting that elementary fermions fall into two similar groups: those which appear to be individually observable (leptons) and those which are not individually observable (quarks).

We now turn to a consideration of a vierbein model of gravity which has certain close similarities to the pseudoquantum field theories we have been studying. In Weyl's formulation[13] of the Einstein-Cartan theory of gravity a vierbein field, $l^{\mu a}(x)$, is introduced which is the "square root" of the metric tensor

$$g^{\mu\nu} = \eta_{ab} l^{\mu a} l^{\nu b}, \tag{169}$$

where η_{ab} is the constant metric tensor of special relativity, where Roman indices transform as vectors under the SL(2, C) group of local Lorentz transformations, and where Greek indices transform as vectors under general coordinate transformations. It is useful to introduce the constant Dirac matrices, γ_a and $4S_{ab} = i[\gamma_a, \gamma_b]$. Under an SL(2, C) transformation,

$$S = \exp[iC^{ab}(x)S_{ab}], \tag{170}$$

a spinor, $\psi(x)$, becomes

$$\psi' = S\psi. \tag{171}$$

The local nature of the transformation requires the introduction of a gauge field

$$B^{ab}_\mu = -B^{ba}_\mu \tag{172}$$

which transforms inhomogeneously,

$$B_\mu \to SB_\mu S^{-1} - \frac{i}{g} S\partial_\mu S^{-1}, \tag{173}$$

so that a Lorentz transformation gauge-covariant derivative can be defined

$$\nabla_\mu \psi = (\partial_\mu + igB_\mu)\psi, \tag{174}$$

where $B_\mu = B^{ab}_\mu S_{ab}$ and $g = 12\pi G$ where G is Newton's constant. Under a gauge transformation we have

$$l^\mu = l^{\mu a}\gamma_a \to Sl^\mu S^{-1}, \tag{175}$$

so that the gauge-covariant derivative of l^μ is defined to be

$$\nabla_\nu l^\mu = (\partial_\nu + igB_\nu \times) l^\mu, \tag{176}$$

where $B_\nu \times l^\mu = [B_\nu, l^\mu]$. The commutator

$$igB_{\mu\nu} = [\partial_\mu + igB_\mu, \partial_\nu + igB_\nu] \tag{177}$$

transforms homogeneously under a gauge transformation

$$B_{\mu\nu} \to SB_{\mu\nu}S^{-1}, \tag{178}$$

and as a second-rank tensor under general coordinate transformations. With these field quantities we are able to construct a Lagrangian $\mathcal{L}_{\text{Weyl}}$ which reduces to the Einstein Lagrangian for gravity when no matter is present,[13]

$$\mathcal{L} = \mathcal{L}_{\text{Weyl}} + \mathcal{L}_{\text{matter}}, \tag{179}$$

where

$$\mathcal{L}_{\text{Weyl}} = \frac{i}{8l} \operatorname{Tr} l^\mu l^\nu B_{\mu\nu} \tag{180}$$

and where, for example, we might let

$$l\,\mathcal{L}_{\text{matter}} = \bar{\psi}(il^\mu \nabla_\mu + m)\psi \tag{181}$$

with $l = \det(l^{\mu a})$.

We observe that the terms containing derivatives in $\mathcal{L}_{\text{Weyl}}$ are linear in the field B_μ—a suggestive feature in view of our previous discussion. However, the quadratic term in B_μ eliminates the possibility of regarding $\mathcal{L}_{\text{Weyl}}$ as a pseudoquantum field theory for a classical field $l^{\mu a}$. But, regardless of this consideration, the fact that $l^{\mu a}$ is necessarily classical in part leads us to consider quantizing vierbein gravity in a manner which is based on the pseudoquantization procedure described above. Remembering that a successful perturbation theory requires the perturbation to be around known solutions we introduce a quadratic Lagrangian term via

$$\mathcal{L} = \mathcal{L}_0 + (\mathcal{L} - \mathcal{L}_0) = \mathcal{L}_0 + \mathcal{L}_{\text{int}}, \tag{182}$$

where

$$\mathcal{L}_0 = -\tfrac{1}{4} i \operatorname{Tr}(B'_{\mu a} l^\mu \gamma^a + ig[B_a, B_b]\gamma^a\gamma^b) \tag{183}$$

and

$$B'_{\mu a}=\partial_\mu B_a-\partial_a B_\mu\,. \tag{184}$$

Our plan is to follow the pseudoquantization procedure for the "free" part of the Lagrangian $\mathcal{L}_0$. Therefore we will (i) choose a particular coordinate system (harmonic coordinates) and a particular gauge, the "Lorentz" gauge, $\partial^\mu B_\mu=0$, (ii) establish equal-time commutation relations, (iii) define a set of eigenstates of $l^{\mu a}$, and (iv) proceed to calculate quantum corrections in perturbation theory.

The equations of motion for the "free" Lagrangian $\mathcal{L}_0$ are

$$\partial_\mu B_b^{ab}-\partial_b B_\mu^{ab}=0 \tag{185}$$

and

$$\partial_\mu(l^{\mu a}\eta^{\nu b}-l^{\nu a}\eta^{\mu b})+2g\,(\eta^{\nu a}B_c^{cb}-\eta^{\nu b}B_c^{ca}-\eta^{ac}B_c^{\nu b}+\eta^{bc}B_c^{\nu a})=0\,. \tag{186}$$

We work in the gravitational equivalent of the Lorentz gauge of electrodynamics,

$$\partial^\mu B_\mu^{ab}=0\,, \tag{187}$$

and choose harmonic coordinates

$$\partial_\mu l^{\mu a}=\tfrac{1}{2}\,\partial^a\eta_{\sigma\tau}l^{\sigma\tau}\,. \tag{188}$$

The Green's function associated with Eq. (185) is

$$G_{\alpha ef,\rho\sigma}(x,y)=-\tfrac{1}{2}\int\frac{d^4k}{k^2}\,e^{-ik\cdot(x-y)}g_{\alpha ef,\rho\sigma}(k)\,, \tag{189}$$

where

$$g_{\alpha ef,\rho\sigma}(k)=k_e\left(\eta_{\alpha\rho}\eta_{f\sigma}+\eta_{\alpha\sigma}\eta_{f\rho}-\eta_{\alpha f}\eta_{\rho\sigma}-\frac{k_\alpha k_\rho\eta_{f\sigma}+k_\alpha k_\sigma\eta_{f\rho}}{k^2}\right)-k_f\left(\eta_{\alpha\rho}\eta_{e\sigma}+\eta_{\alpha\sigma}\eta_{e\rho}-\eta_{\alpha e}\eta_{\rho\sigma}-\frac{k_\alpha k_\rho\eta_{e\sigma}+k_\alpha k_\sigma\eta_{e\rho}}{k^2}\right). \tag{190}$$

In order to relate the above Green's function to a time-ordered product of the quantum fields it is first necessary to introduce a set of coherent states, $|L\rangle$, which are eigenstates of $l^{\mu a}$:

$$l^{\mu a}(x)\,|L\rangle=L^{\mu a}(x)\,|L\rangle\,, \tag{191}$$

where $L^{\mu a}(x)$ is a c-number function of x. In particular, we define $|\eta\rangle$ to satisfy

$$l^{\mu a}|\eta\rangle=\eta^{\mu a}|\eta\rangle\,, \tag{192}$$

where $\eta^{\mu a}$ is the constant Lorentz metric tensor of special relativity. Given a state $|L\rangle$ we define the field

$$l_L^{\mu a}=l^{\mu a}-L^{\mu a}\,. \tag{193}$$

This field corresponds to the quantum part of $l^{\mu a}$ and when applied to the purely classical state $|L\rangle$ has the eigenvalue zero.

We now make the identification

$$iG_{\alpha ef,\rho\sigma}(x,y)=\langle L\,|\,T(B_{\alpha ef}(x),l_{L\rho\sigma}(y))\,|\,L\rangle\,. \tag{194}$$

If we desire to calculate quantum corrections to $l_{\rho\sigma}=\eta_{\rho\sigma}$ we choose $|L\rangle=|\eta\rangle$. (It should be noted that $G_{\alpha ef,\rho\sigma}$ is independent of the choice of $|L\rangle$ as we have defined it.) Because $l_{L\rho\sigma}(y)$ is sharp on $|L\rangle$ we find that the right side of Eq. (194) becomes

$$iG_{\alpha ef,\rho\sigma}(x,y)=\theta(y_0-x_0)[l_{\rho\sigma}(y),B_{\alpha ef}(x)] \tag{195}$$

up to a functional δ function. From the form of $\mathcal{L}_0$ we see that the commutator is not zero. It is fully determined by an equal-time commutation relation of $l_{\rho\sigma}$ and $B_{\alpha ef}$ (which by the way is the only nonzero equal-time commutator if the canonical procedure is followed), the equations of motion, and the requirement that it be zero at spacelike distances. The "retarded" form of $G_{\alpha ef,\rho\sigma}$ fixes the integration contour around poles in Eq. (192). The other nonzero Green's function in the free Lagrangian model specified by $\mathcal{L}_0$ is

$$iH^{\mu\nu,\rho\sigma}(x,y)=\langle L\,|\,T(l_L^{\mu\nu}(x),l_L^{\rho\sigma}(y))\,|\,L\rangle\,. \tag{196}$$

It is nonzero owing to the presence of the $[B_\mu,B_\nu]$ term in $\mathcal{L}_0$. We shall show in the next section that it is a principal-value propagator rather than a Feynman propagator. In coordinate space this results in $H^{\mu\nu,\rho\sigma}$ being the sum of the advanced and retarded propagators. As a result our model is equivalent to an action-at-a-distance theory in some sectors.

The classical part of $l_{\mu a}$ is the solution of the classical linearized field equations with appropriate matter sources. The linearized field equations are derived from a Lagrangian consisting of $\mathcal{L}_0$ plus matter terms. (Note that the form of $\mathcal{L}_0$ is obtained by substituting $l_{\mu a}=\eta_{\mu a}+h_{\mu a}$ in $\mathcal{L}_{\text{Weyl}}$, expanding, and keeping quadratic terms.) Thus the class of possible background metrics is restricted.

A simplification occurs in perturbation theory when the classical part of $l_{\mu\alpha}$ is $\eta_{\mu\alpha}$. In this case $(\mathcal{L}_{\text{Weyl}}-\mathcal{L}_0)|\eta\rangle=0$ when $\mathcal{L}_0$ and $\mathcal{L}_{\text{Weyl}}$ are expressed in terms of asymptotic fields.

V. PRINCIPAL-VALUE PROPAGATORS AND ACTION AT A DISTANCE

In this section we shall show that certain propagators, in field theories where the pseudoquantization procedure has been followed, are principal-value propagators (i.e., the sum of the advanced and retarded Green's functions in coordinate space) rather than Feynman propagators. We also describe a quantum field theory for action-at-a-distance electrodynamics which completes the program initiated by Schwarzschild, Tetrode, and Fokker.[14]

To illustrate the origin of the principal-value propagator we return to the scalar field model of Eq. (103) which described a classical field, $\phi_1(x)$. We introduce an interaction term

$$L_{\rm int} = -\int d^3z\, \tfrac{1}{2}\lambda^2\, [\phi_2(z)]^2 \tag{197}$$

(where λ is a constant), which destroys the purely classical nature of ϕ_1. Suppose we consider the Green's function

$$i\bar{G}(x,y) = \langle 0|\, T(\phi_1(x)\phi_1(y))\,|0\rangle\,, \tag{198}$$

which would be zero if $L_{\rm int}$ were not present. In terms of in-fields we have

$$i\bar{G}(x,y) = \left\langle 0 \left| T\left(\phi_{1\rm in}(x)\phi_{1\rm in}(y)\exp\left(i\int dt\, L_{\rm int}\right)\right)\right| 0\right\rangle, \tag{199}$$

where the vacuum states, $|0\rangle$ and $\langle 0|$, are defined as in Eqs. (120) and (122). From the definition of the vacuum we find (dropping "in" labels)

$$i\bar{G}(x,y) = \frac{-i\lambda^2}{2}\int d^4z\langle 0|\, T(\phi_1(x)\phi_1(y)\phi_2{}^2(z))\,|0\rangle\,, \tag{200}$$

which becomes

$$i\bar{G}(x,y) = \frac{-i\lambda^2}{2}\,\epsilon(x_0-y_0)\,\frac{\partial}{\partial m^2}\,\Delta(x-y) \tag{201}$$

with

$$\Delta(x-y) = -i\int \frac{d^4k}{(2\pi)^3}\,\delta(k^2-m^2)\epsilon(k_0)e^{-ik\cdot(x-y)}\,. \tag{202}$$

Using

$$\tfrac{1}{2}\epsilon(x_0-y_0)\Delta(x-y) = \int \frac{d^4k}{(2\pi)^4}\,\mathrm{P}\,\frac{1}{k^2-m^2} \times e^{-ik\cdot(x-y)}\,, \tag{203}$$

we see that

$$\bar{G}(x,y) = -\lambda^2\int \frac{d^4k}{(2\pi)^4}\,\mathrm{P}\,\frac{1}{(k^2-m^2)^2}\,e^{-ik\cdot(x-y)}\,, \tag{204}$$

where

$$\mathrm{P}\,\frac{1}{(k^2-m^2)^2} = \frac{1}{2}\left[\frac{1}{(k^2-m^2+i\epsilon)^2} + \frac{1}{(k^2-m^2-i\epsilon)^2}\right]. \tag{205}$$

The form of $\bar{G}$ is consistent with the equations of motion:

$$(\Box + m^2)\phi_1 + \lambda^2\phi_2 = 0\,, \tag{206}$$

$$(\Box + m^2)\phi_2 = \delta^4(x-y)\,. \tag{207}$$

The appearance of the principal-value dipole propagator rather than the Feynman dipole propagator in Eq. (204) is useful because it eliminates certain unitarity problems associated with indefinite-metric fields. However, depending on the model under consideration, it could lead to difficulties with causality. To illustrate the manner in which unitarity problems are resolved, consider the interaction of the ϕ_1 dipole field with a scalar quantum field ψ with

$$L'_{\rm int} = g\phi_1(x)[\psi(x)]^2\,. \tag{208}$$

Suppose we consider the subset of in and out states containing arbitrary numbers of ψ particles but no ϕ_1 or ϕ_2 particles. These states have positive metric. If one could systematically exclude indefinite-metric ϕ_1 and ϕ_2 particles from physical states one would avoid negative probabilities and other problems. But the sum over states in a unitarity sum would normally include states with ϕ_1 particles if the ϕ_1 field had Feynman propagators. In the case of principal-value propagators, no intermediate states with ϕ_1 particles occur, since the pole term is not present. The interaction mediated by the ϕ_1 field is a form of action at a distance and ϕ_1 is properly described by the phrase adjunct field, coined by Feynman and Wheeler.[14] A more detailed discussion of the unitarity question is given in Refs. 7 and 8. In those articles a dipole gluon model for quark confinement was proposed which introduced principal-value propagators in an *ad hoc* manner to resolve unitarity problems. It was pointed out that causality problems did not necessarily exist in those models because the non-Abelian dipole gluons were confined for the same reason as the quarks so that—at the worst—there would be unobservable causality violations at distances of the order of hadron dimensions.

The pseudoquantization procedure may be used to construct a quantum field-theoretic version of action-at-a-distance electrodynamics. Consider the Lagrangian

$$\mathcal{L} = -\tfrac{1}{2}F^{\mu\nu}(\partial_\nu A_\mu - \partial_\mu A_\nu) + \tfrac{1}{4}F^{\mu\nu}F_{\mu\nu} + \bar{\psi}(i\not{\partial} - e\not{A} - m_0)\psi\,. \tag{209}$$

We define the momentum

$$\Pi_\mu = \frac{\delta \mathfrak{L}}{\delta \dot{A}^\mu} = F_{0\mu} . \tag{210}$$

Going to the transverse gauge as in Sec. IV, we define the equal-time commutation relation

$$[\Pi_i(\vec{x}, t), A_j(\vec{y}, t)] = i\delta_{ij}^{tr}(\vec{x} - \vec{y}) . \tag{211}$$

Suppose we neglect interaction terms in $\mathfrak{L}$ for the moment and choose $F_{\mu\nu}$ to be an observable classical field (as it is up to quantum corrections which we neglect) and A_μ to be unobservable (as it is because it is not gauge invariant). Then we follow our pseudoquantization procedure for

$$\mathfrak{L}_0 = -\tfrac{1}{2} F^{\mu\nu}(\partial_\nu A_\mu - \partial_\mu A_\nu) + \tfrac{1}{4} F^{\mu\nu} F_{\mu\nu} . \tag{212}$$

In particular, we define a vacuum such that

$$F_{\mu\nu}|0\rangle = 0, \quad A_\mu|0\rangle \neq 0 , \tag{213}$$

while

$$\langle 0|A_\mu = 0, \quad \langle 0|F_{\mu\nu} \neq 0 . \tag{214}$$

Then

$$iG_{\mu\nu}(x, y) = \langle 0| T(A_\mu(x) A_\nu(y))|0\rangle \tag{215}$$

would be zero were it not for $F_{\mu\nu}F^{\mu\nu}$ in $\mathfrak{L}_0$. In terms of appropriate in-fields it becomes

$$2iG_{\mu\nu}(x, y) = \int d^4z\, (\theta(x_0 - y_0)\theta(y_0 - z_0) + \theta(y_0 - x_0)\theta(x_0 - z_0)) \times [A_{\mu\,\text{in}}(x), F_{\alpha\beta\,\text{in}}(z)][A_{\mu\,\text{in}}(y), F^{\alpha\beta}_{\text{in}}(z)] . \tag{216}$$

Note that we are treating $F_{\mu\nu}F^{\mu\nu}$ in $\mathfrak{L}_0$ as an interaction term. The structure of $G_{\mu\nu}(x, y)$ is the same as that of Eq. (200) so we can conclude that

$$G_{\mu\nu}(x, y) = -g_{\mu\nu} \int \frac{d^4k}{(2\pi)^4}\, \mathrm{P}\frac{1}{k^2}\, e^{-ik\cdot(x-y)} \tag{217}$$

in the Feynman gauge. Thus the action-at-a-distance interaction follows from the pseudoquantization of electrodynamics. The classical character of $F_{\mu\nu}$ is lost owing to quantum corrections resulting from the presence of $J_\mu A^\mu$ in the Lagrangian.

The example we have just studied has a certain parallel in the vierbein model of gravitation studied in the last section. The forms of the Lagrangian and commutation relations are similar. As a result it is clear that

$$D^{\mu\nu,\lambda\sigma}(x, y) = \left\langle L \left| T\left(l^{\mu\nu}_{L\,\text{in}}(x) l^{\lambda\sigma}_{L\,\text{in}}(y) \int d^4z\, \bar{\mathfrak{L}}_{\text{int}}(z)\right) \right| L \right\rangle \tag{218}$$

with

$$\bar{\mathfrak{L}}_{\text{int}} = \tfrac{1}{4} g \operatorname{Tr}[B_{\mu\,\text{in}}, B_{\nu\,\text{in}}]\gamma^\mu\gamma^\nu \tag{219}$$

is a principal-value propagator. Therefore we have constructed an action-at-a-distance version of quantum gravity. Our motivation was to take account of the classical part of $l^{\mu a}$ in a way which did not divorce it from the quantum part to which it is intimately related.

VI. CONCLUSION

We have seen that an alternative to Fock-space quantization exists for a class of field theories which have Lagrangian gradient terms which are linear in field variables. A method was also proposed for constructing Lagrangians of that type from classical Lagrangians with gradient terms which are quadratic in field variables. To some extent this process has a parallel in the passage from Klein-Gordon field Lagrangians which are quadratic in derivatives to Dirac field Lagrangians which are linear in derivatives.

The quantization procedure we have outlined is canonical so far as the fields are concerned. We do, however, make a choice of vacuum states which differs from the usual choice. As a result we have found free propagators which were either retarded, or half-advanced and half-retarded. The choice of vacuum state does not in itself preclude the appearance of Feynman propagators. If one has a good reason to modify the canonical commutation relations then it is possible to obtain Feynman propagators.[15] The procedure we have outlined has, therefore, a greater generality than the particular class of models studied in the present work. It can enable one to embed a classical field theory in a quantum field theory in such a way as to maintain its classical character. It can also be applied to study classical field theories which obtain quantum corrections. Finally it can be applied in order to obtain a fully second-quantized field theory (cf. Ref. 15).

ACKNOWLEDGMENT

This work was supported in part by the U.S. Energy Research and Development Administration.

*Present address: Physics Department, Williams College, Williamstown, Mass. 01267.

[1]D. R. Yennie, S. C. Frautschi, and H. Suura, Ann. Phys. (N.Y.) 13, 379 (1961).

[2]R. J. Glauber, Phys. Rev. 131, 2766 (1963).

[3]W. A. Bardeen, M. S. Chanowitz, S. D. Drell, M. Weinstein, and T.-M. Yan, Phys. Rev. D 11, 1094 (1975).

[4]J. M. Cornwall and G. Tiktopoulos, Phys. Rev. D 13, 3370 (1976).
[5]S. Blaha, Phys. Lett. 56B, 373 (1975).
[6]E. C. G. Sudarshan, Center for Particle Theory report Univ. of Texas—Austin, 1976 (unpublished).
[7]S. Blaha, Phys. Rev. D 10, 4268 (1974).
[8]S. Blaha, Phys. Rev. D 11, 2921 (1975).
[9]S. Blaha, Lett. Nuovo Cimento 18, 60 (1977).
[10]Cf. Ref. 2; T. W. B. Kibble, J. Math. Phys. 9, 315 (1968); Phys. Rev. 173, 1527 (1968), 174, 1882 (1968), 175, 1624 (1968);
[11]A. Salam, lecture at Center for Theoretical Studies, Miami, Florida, 1973 (unpublished).
[12]G. Mie, Ann. Phys. (Leipzig) 37, 511 (1912); 39, 1 (1912); 40, 1 (1913); H. Weyl, *Space, Time, Matter* (Dover, N.Y. 1952).
[13]H. Weyl, Z. Phys. 56, 330 (1929); T. W. B. Kibble, J. Math. Phys. 2, 212 (1961); J. Schwinger, Phys. Rev. 130, 1253 (1963); C. J. Isham, A. Salam, and J. Strathdee, Lett. Nuovo Cimento 5, 969 (1972); F. W. Hehl, P. von der Heyde, G. D. Kerlick, and J. Nester, Rev. Mod. Phys. 48, 393 (1976); and references therein.
[14]K. Schwarzschild, Göttinger Nachrichten 128, 132 (1903); H. Tetrode, Z. Phys. 10, 317 (1922); A. D. Fokker, *ibid.* 58, 386 (1929); J. Wheeler and R. P. Feynman, Rev. Mod. Phys. 17, 157 (1945); 21, 425 (1949).
[15]S. Blaha (unpublished).

REFERENCES

Bjorken, J. D., Drell, S. D., 1964, *Relativistic Quantum Mechanics* (McGraw-Hill, New York, 1965).

Bjorken, J. D., Drell, S. D., 1965, *Relativistic Quantum Fields* (McGraw-Hill, New York, 1965).

Blaha, S., 1998, *Cosmos and Consciousness* (Pingree-Hill Publishing, Auburn, NH, 1998).

_______, 2002, *A Finite Unified Quantum Field Theory of the Elementary Particle Standard Model and Quantum Gravity Based on New Quantum Dimensions™ & a New Paradigm in the Calculus of Variations* (Pingree-Hill Publishing, Auburn, NH, 2002).

_______, 2003, *A Finite Unified Quantum Field Theory of the Elementary Particle Standard Model and Quantum Gravity Based on New Quantum Dimensions™ and a New Paradigm in the Calculus of Variations* (Pingree-Hill Publishing, Auburn, NH, 2003).

_______, 2004, *Quantum Big Bang Cosmology: Complex Space-time General Relativity, Quantum Coordinates™ Dodecahedral Universe, Inflation, and New Spin 0, ½, 1 & 2 Tachyons & Imagyons* (Pingree-Hill Publishing, Auburn, NH, 2004).

______, *2005a, Quantum Theory of the Third Kind: A New Type of Divergence-free Quantum Field Theory Supporting a Unified Standard Model of Elementary Particles and Quantum Gravity based on a New Method in the Calculus of Variations* (Pingree-Hill Publishing, Auburn, NH, 2005).

______, 2005b, *The Metatheory of Physics Theories, and the Theory of Everything as a Quantum Computer Language* (Pingree-Hill Publishing, Auburn, NH, 2005).

______, 2005c, *The Equivalence of Elementary Particle Theories and Computer Languages: Quantum Computers, Turing Machines, Standard Model, Superstring Theory, and a Proof that Gödel's Theorem Implies Nature Must Be Quantum* (Pingree-Hill Publishing, Auburn, NH, 2005).

______, 2006a, *The Foundation of the Forces of Nature* (Pingree-Hill Publishing, Auburn, NH, 2006).

______, 2006b, *A Derivation of ElectroWeak Theory based on an Extension of Special Relativity; Black Hole Tachyons; & Tachyons of Any Spin.* (Pingree-Hill Publishing, Auburn, NH, 2006).

_______, 2007a, *Physics Beyond the Light Barrier: The Source of Parity Violation, Tachyons, and A Derivation of Standard Model Features* (Pingree-Hill Publishing, Auburn, NH, 2007).

______, 2007b, *The Origin of the Standard Model: The Genesis of Four Quark and Lepton Species, Parity Violation, the ElectroWeak Sector, Color SU(3), Three Visible Generations of Fermions, and One Generation of Dark Matter with Dark Energy* (Pingree-Hill Publishing, Auburn, NH, 2007).

______, *2008a, A Direct Derivation of the Form of the Standard Model From GL(16) (Pingree-Hill Publishing, Auburn, NH, 2008).*

______, 2008b, *A Complete Derivation of the Form of the Standard Model With a New Method to Generate Particle Masses Second Edition* (Pingree-Hill Publishing, Auburn, NH, 2008)

______, 2009, *The Algebra of Thought & Reality: The Mathematical Basis for Plato's Theory of Ideas, and Reality Extended to Include A Priori Observers and Space-Time Second Edition* (Pingree-Hill Publishing, Auburn, NH, 2009).

______, 2010a, *Operator Metaphysics: A New Metaphysics Based on a New Operator Logic and a New Quantum Operator Logic that Lead to a Mathematical Basis for Plato's Theory of Ideas and Reality* (Pingree-Hill Publishing, Auburn, NH, 2010).

______, 2010b, *The Standard Model's Form Derived from Operator Logic, Superluminal Transformations and GL(16)* (Pingree-Hill Publishing, Auburn, NH, 2010).

______, 2011a, *21st Century Natural Philosophy Of Ultimate Physical Reality* (McMann-Fisher Publishing, Auburn, NH, 2011).

______, 2011b, *All the Universe! Faster Than Light Tachyon Quark Starships & Particle Accelerators with the LHC as a Prototype Starship Drive Scientific Edition* (Pingree-Hill Publishing, Auburn, NH, 2011).

______, 2011c, *From Asynchronous Logic to The Standard Model to Superflight to the Stars* (Blaha Research, Auburn, NH, 2011).

______, 2012a, *From Asynchronous Logic to The Standard Model to Superflight to the Stars volume 2: Superluminal CP and CPT, U(4) Complex General Relativity and The Standard Model, Complex Vierbein General Relativity, Kinetic Theory, Thermodynamics* (Blaha Research, Auburn, NH, 2012).

______, 2012b, *Standard Model Symmetries, And Four And Sixteen Dimension Complex Relativity; The Origin Of Higgs Mass Terms* (Blaha Reasearch, Auburn, NH, 2012).

______, 2013a, *Multi-Stage Space Guns, Micro-Pulse Nuclear Rockets, and Faster-Than-Light Quark-Gluon Ion Drive Starships* (Blaha Research, Auburn, NH, 2013).

______, 2013b, *The Bridge to Dark Matter; A New Sister Universe; Dark Energy; Inflatons; Quantum Big Bang; Superluminal Physics; An Extended Standard Model Based on Geometry* (Blaha Reasearch, Auburn, NH, 2013).

______, 2014a, *Universes and Multiverses: From a New Standard Model to a Physical Multiverse; The Big Bang; Our Sister Universe's Wormhole; Origin of the Cosmological Constant, Spatial Asymmetry of the Universe, and its Web of Galaxies; A Baryonic Field between Universes and Particles; Flatverse Extended Wheeler-DeWitt Equation* (Blaha Reasearch, Auburn, NH, 2014).

______, 2014b, *All the Multiverse! Starships Exploring the Endless Universes of the Cosmos Using the Baryonic Force* (Blaha Research, Auburn, NH, 2014).

______, 2014c, *All the Multiverse! II Between Multiverse Universes: Quantum Entanglement Explained by the Multiverse Coherent Baryonic Radiation Devices – PHASERs Neutron Star Multiverse Slingshot Dynamics Spiritual and UFO Events, and the Multiverse Microscopic Entry into the Multiverse* (Blaha Research, Auburn, NH, 2014).

______, 2015a, *PHYSICS IS LOGIC PAINTED ON THE VOID: Origin of Bare Masses and The Standard Model in Logic, U(4) Origin of the Generations, Normal and Dark Baryonic Forces, Dark Matter, Dark Energy, The Big Bang, Complex General Relativity, A Megaverse of Universe Particles* (Blaha Research, Auburn, NH, 2015).

______, 2015b, *PHYSICS IS LOGIC Part II: The Theory of Everything, The Megaverse Theory of Everything, U(4)⊗U(4) Grand Unified Theory (GUT), Inertial Mass = Gravitational Mass, Unified Extended Standard Model and a New Complex General Relativity with Higgs Particles, Generation Group Higgs Particles* (Blaha Research, Auburn, NH, 2015).

Eddington, A. S., 1952, *The Mathematical Theory of Relativity* (Cambridge University Press, Cambridge, U.K., 1952).

Fant, Karl M., 2005, *Logically Determined Design: Clockless System Design With NULL Convention Logic* (John Wiley and Sons, Hoboken, NJ, 2005).

Gelfand, I. M., Fomin, S. V., Silverman, R. A. (tr), 2000, *Calculus of Variations* (Dover Publications, Mineola, NY, 2000).

Heitler, W., 1954, *The Quantum Theory of Radiation* (Claendon Press, Oxford, UK, 1954).

Huang, Kerson, 1992, *Quarks, Leptons & Gauge Fields 2nd Edition* (World Scientific Publishing Company, Singapore, 1992).

Jost, J., Li-Jost, X., 1998, *Calculus of Variations* (Cambridge University Press, New York, 1998).

Rescher, N., 1967, *The Philosophy of Leibniz* (Prentice-Hall, Englewood Cliffs, NJ, 1967).

Sagan, H., 1993, *Introduction to the Calculus of Variations* (Dover Publications, Mineola, NY, 1993).

Sakurai, J. J., 1964, *Invariance Principles and Elementary Particles* (Princeton University Press, Princeton, NJ, 1964).

Streater, R. F. and Wightman, A. S., 2000, *PCT, Spin, Statistics, and All That* (Princeton University Press, Princeton, NJ 2000).

Weinberg, S., 1972, *Gravitation and Cosmology* (Joh Wiley and Sons, New York, 1972).

Weinberg, S., 1995, *The Quantum Theory of Fields Volume I* (Cambridge University Press, New York, 1995).

Weyl, H., 1950, *Space, Time, Matter* (Dover, New York, 1950).

Weyl, H., (Tr. S. Pollard et al), 1987, *The Continuum* (Dover Publications, New York, 1987).

INDEX

About the Author

Stephen Blaha is a well known Physicist and Man of Letters with interests in Science, Society and civilization, the Arts, and Technology. He had an Alfred P. Sloan Foundation scholarship in college. He received his Ph.D. in Physics from Rockefeller University. He has served on the faculties of several major universities. He was also a Member of the Technical Staff at Bell Laboratories, a manager at the Boston Globe Newspaper, a Director at Wang Laboratories, and President of Blaha Software Inc and of Janus Associates Inc. (NH).

Among other achievements he was a co-discoverer of the "r potential" for heavy quark binding developing the first (and still the only demonstrable) non-abelian gauge theory with an "r" potential; first suggested the existence of topological structures in superfluid He-3; first proposed Yang-Mills theories would appear in condensed matter phenomena with non-scalar order parameters; first developed a grammar-based formalism for quantum computers and applied it to elementary particle theories; first developed a new form of quantum field theory without divergences (thus solving a major 60 year old problem that enabled a unified theory of the Standard Model and Quantum Gravity without divergences to be developed); first developed a formulation of complex General Relativity based on analytic continuation from real space-time; first developed a generalized non-homogeneous Robertson-Walker metric that enabled a quantum theory of the Big Bang to be developed without singularities at $t = 0$; first generalized Cauchy's theorem and Gauss' theorem to complex, curved multi-dimensional spaces; received Honorable Mention in the Gravity Research Foundation Essay Competition in 1978; first developed a physically acceptable theory of faster-than-light particles; first derived a composition of extrema method in the Calculus of Variations; first quantitatively suggested that inflationary periods in the history of the universe were not needed; first proved Gödel's Theorem implies Nature must be quantum; provided a new alternative to the Higgs Mechanism, and Higgs particles, to generate masses; first showed how to resolve logical paradoxes including Gödel's Undecidability Theorem by developing Operator Logic and Quantum Operator Logic; first developed a quantitative harmonic oscillator-like model of the life cycle, and interactions, of civilizations; first showed how equations describing superorganisms also apply to civilizations. A recent book shows his theory applies successfully to the past 14 years of history and to *new* archaeological data on Andean and Mayan civilizations as well as Early Anatolian and Egyptian civilizations.

He first developed an axiomatic derivation of the forms of The Standard Model from geometry – space-time properties – The Extended Standard Model. It has a Dark Matter sector that approximates the ElectroWeak sector with Dark doublets and Dark gauge interactions. It also uses quantum coordinates to remove infinities that crop up in most interacting quantum field theories and additionally to remove the infinities that appear in the Big Bang and generate an inflationary growth of the universe. The Extended Standard Model has an ultra-high energy GUT (Grand Unified Theory) limit with a $U(4) \otimes U(4)$ symmetry; and can be united with gravitation to form a Theory of Everything. (See *Physics is Logic Part II.*)

Blaha has had a major impact on a succession of elementary particle theories: his Ph.D. thesis (1970), and papers, showed that quantum field theory calculations to all orders in ladder approximations could not give scaling deep inelastic electron-nucleon scattering. He later showed the eigenvalue equation for the fine structure constant α in Johnson-Baker-Willey QED had a zero at $\alpha = 1$ not 1/137 by solving the Schwinger-Dyson equations to all orders in an approximation that agreed with exact results to 4th order in α thus ending interest in this theory. In 1979 at Prof. Ken Johnson's (MIT) suggestion he calculated the proton-neutron mass difference in the MIT bag model and found the result had the wrong sign reducing interest in the bag model. These results all appear in Physical Review papers. In the 2000's he repeatedly pointed out the shortcomings of SuperString theory and showed that The Standard Model's form could be derived from space-time geometry by an extension of Lorentz transformations to faster than light transformations. This deeper space-time basis greatly increases the possibility that it is part of THE fundamental theory.

In graduate school (1965-71) he wrote substantial papers in elementary particles and group theory: The Inelastic E- P Structure Functions in a Gluon Model. Phys. Lett. B40:501-502,1972; Deep-Inelastic E-P Structure Functions In A Ladder Model With Spin 1/2 Nucleons, Phys.Rev. D3:510-523,1971; Continuum Contributions To The Pion Radius, Phys. Rev. 178:2167-2169,1969; Character Analysis of U(N) and SU(N), J. Math. Phys. 10, 2156 (1969); and The Calculation of the Irreducible Characters of the Symmetric Group in Terms of the Compound Characters, (Published as Blaha's Lemma in D. E. Knuth's book: *The Art of Computer Programming Vols. 1 – 4*).

In the early 1980's Blaha was also a pioneer in the development of UNIX for financial, scientific and Internet applications: benchmarked UNIX versions showing that block size was critical for UNIX performance, developing financial modeling software, starting database benchmarking comparison studies, developing Internet-like UNIX networking (1982) and developing a hybrid shell programming technique (1982) that was a precursor to the PERL programming language. He was also the manager of the AT&T ten-year future products development database. His work helped lead to commercial UNIX on computers such as Sun Micros, IBM AIX minis, and Apple computers.

In the 1980's he pioneered the development of PC Desktop Publishing on laser printers. and was nominated for three "Awards for Technical Excellence" in 1987 by PC Magazine for PC software products that he designed and developed.

Recently he has developed a theory of Megaverses – actual universes of which our universe is one – with quantum particle-like properties based on the Wheeler-DeWitt equation of Quantum Gravity. He has developed a theory of a baryonic force, which had been conjectured many years ago, and estimated the strength of the force based on discrepancies in measurements of the gravitational constant G. This force, operative in

15-dimensinal space, can be used to escape from our universe in "uniships" which are the equivalent of the faster-than-light starships proposed in the author's earlier books. Thus travel to other universes, as well as to other stars is possible.

Blaha also considered the complexified Wheeler-DeWitt equation and showed that its limitation to real-valued coordinates and metrics generated a Cosmological Constant in the Einstein equations.

The author has also recently written a series of books on the serious problems of the United States and their solution as well as a book on the decline of Mankind that will follow from current social and genetic trends in Mankind.

In the past twelve years Dr. Blaha has written over 40 books on a wide range of topics. Some recent major works are: *From Asynchronous Logic to The Standard Model to Superflight to the Stars, All the Universe!, SuperCivilizations: Civilizations as Superorganisms, America's Future: an Islamic Surge, ISIS, al Qaeda, World Epidemics, Ukraine, Russia-China Pact, US Leadership Crisis,The Rises and Falls of Man – Destiny –*

3000 AD: New Support for a Superorganism MACRO-THEORY of CIVILIZATIONS From CURRENT WORLD TRENDS and NEW Peruvian, Pre-Mayan, Mayan, Anatolian, and Early Egyptian Data, with a Projection to 3000 AD, and *Mankind in Decline: Genetic Disasters, Human-Animal Hybrids, Overpopulation, Pollution, Global Warming, Food and Water Shortages, Desertification, Poverty, Rising Violence, Genocide, Epidemics, Wars, Leadership Failure.*

He has taught approximately 4,000 students in undergraduate, graduate, and postgraduate corporate education courses primarily in major universities, and large companies and government agencies.

The above paragraphs summarize much of his work over the past fifty years. This work is fully documented. He continues to engage in research and writing at Blaha Research.

www.ingramcontent.com/pod-product-compliance
Lightning Source LLC
Chambersburg PA
CBHW082009190326
41458CB00010B/3134